CITIES AND FASCINATION

Re-materialising Cultural Geography

Dr Mark Boyle, Department of Geography, University of Strathclyde, UK,
Professor Donald Mitchell, Maxwell School, Syracuse University, USA and
Dr David Pinder, Queen Mary University of London, UK

Nearly 25 years have elapsed since Peter Jackson's seminal call to integrate cultural geography back into the heart of social geography. During this time, a wealth of research has been published which has improved our understanding of how culture both plays a part in, and in turn, is shaped by social relations based on class, gender, race, ethnicity, nationality, disability, age, sexuality and so on. In spite of the achievements of this mountain of scholarship, the task of grounding culture in its proper social contexts remains in its infancy. This series therefore seeks to promote the continued significance of exploring the dialectical relations which exist between culture, social relations and space and place. Its overall aim is to make a contribution to the consolidation, development and promotion of the ongoing project of re-materialising cultural geography.

Also in the series

Cities and Fascination

Beyond the Surplus of Meaning

Edited by

HEIKO SCHMID
University of Jena, Germany

WOLF-DIETRICH SAHR
Universidade Federal do Paraná, Brasil

and

JOHN URRY
Lancaster University, UK

Routledge
Taylor & Francis Group

LONDON AND NEW YORK

First published 2011 by Ashgate Publishing

Published 2016 by Routledge
2 Park Square, Milton Park, Abingdon, Oxfordshire OX14 4RN
711 Third Avenue, New York, NY 10017, USA

First issued in paperback 2016

Routledge is an imprint of the Taylor & Francis Group, an informa business

British Library Cataloguing in Publication Data
Cities and fascination: beyond the surplus of meaning. – (Re-materialising cultural geography)
   1. City and town life. 2. City dwellers–Psychology. 3. Cultural geography.
   I. Series II. Schmid, Heiko. III. Sahr, Wolf-Dietrich. IV. Urry, John.
   307.1'6-dc22

Library of Congress Cataloging-in-Publication Data
Schmid, Heiko.
   Cities and fascination: beyond the surplus of meaning / by Heiko Schmid, Wolf-Dietrich Sahr and John Urry.
   p. cm. – (Re-materialising cultural geography)
   Includes index.
   ISBN 978-1-4094-1853-5 (hardback: alk. paper)
   1. Sociology, Urban. 2. Cities and towns. 3. City and town life. 4. Urbanization. 5. City planning. I. Sahr, Wolf-Dietrich. II. Urry, John. III. Title.
   HT151.S288 2010
   307.76–dc22

                                               2010034088

ISBN 13: 978-1-138-25509-8 (pbk)
ISBN 13: 978-1-4094-1853-5 (hbk)

Contents

PART IV: CONSEQUENCES OF FASCINATION: NEW HORIZONS

List of Figures and Table

Preface

From the Hanging Gardens of Babylon and Jehovah's Temple of Jerusalem to the boulevards of Haussmann's Paris and the shiny desert cities of Dubai and Las Vegas, large cities and their landmarks have always fascinated people because of their images and symbols. Though the theming of urban landscapes has been a common subject in urban studies since the 1980s, the emotional function of symbolism has surprisingly been largely neglected in the academic research. So even today, most academic traditions are still profoundly influenced by functional, critical, and semiotic approaches, subordinating the subjective and aesthetic dimensions to a mere secondary category. But Georg Simmel's investigation of the psychology of metropolitan man, and the contributions of the critical 'Frankfurt School' have clearly pointed to the fact that capitalist urbanism goes far beyond such a functionalism, and even beyond semiotics.

In this context, a symposium was held in Heidelberg (Germany) from 8–10 November 2007, aiming to add to the discussion with ideas highlighting the different dimensions of fascination in the city as an everyday phenomenon. The symposium 'Economy of Fascination: Themed Urban Landscapes of Postmodernity' was intended to reflect primarily on the role of spectacularization and fascination in the recent transformations of urban landscapes into themed environments. Being hosted in the heart of the historic city of Heidelberg, itself a city with a remarkable 'Romantic' theme, the conference attracted scholars from different countries and disciplines. During the meeting an inspiring debate began on the phenomenon of 'fascination' in and of the city, revealing a great diversity of sometimes controversial positions on the subject. These positions however together formed a network in which almost all the contributions focused on the links between four human dimensions: aesthetics, emotion, experience, and power.

Originally it was the intention of the symposium to publish conventional proceedings. However, the diverse and continuing discussions encouraged the editors to venture into a more conceptual book on 'Cities and Fascination'. We now offer to a broader public the results of these discussions, which have continued during the process of making this book. We hope to hereby give an insight into the subject of fascination in and of the city beyond conventional terms and theories, so that the phenomenon is simultaneously approached from very different perspectives, and where even the organizers interpret the phenomenon from different angles. Thus the book cuts across traditional academic boundaries in the fields of urban studies, architecture, urban sociology, urban geography and cultural studies, but we hope it will fascinate because of its very diversity and the attempt to pave the ground for a difficult dialogue.

In this respect, we would like to thank all those who have contributed to the conference 'Economy of Fascination: Themed Urban Landscapes of Postmodernity' in Heidelberg, and especially for those who later permitted the publication of their contributions after all having reworked the chapters of this book. We thank the *Deutsche Forschungsgemeinschaft* [German Research Foundation] and the Foundation of the University of Heidelberg for providing the funds for hosting this conference at the Internationales Wissenschaftsforum Heidelberg [International Academic Forum Heidelberg]. We are also very grateful to Valerie Rose and Jude Chillman at Ashgate for their support, and to Sabine Heurich for her editorial assistance. Very special thanks go to Peter Bews for his invaluable linguistic expertise, as the book also reunites different language and thought traditions so that his review of the manuscripts has become an adventure in itself.

Jena, Curitiba, Lancaster, December 2010

Acknowledgements

The editors and publishers wish to thank the following for permission to use copyright material:

Chôros Laboratory of the Swiss Federal Institute of Technology, Lausanne (EPFL) for the map of the election results of the Swiss referendum (popular initiative) on the banning of the construction of minarets 2009 (see fig. 3.4).

Lit Verlag for the photo of a suburb in Surrey, Canada (see fig. 5.2).

Studio Hamburg Produktion GmbH for the film still of *Die Entscheidung* [*The Decision*], Germany 2005 (see fig. 10.1).

Element E Filmproduktion for the set photo of *Paulas Geheimnis* [*Paula's Secret*], Germany 2006 (see fig. 10.2).

Hamburg Marketing GmbH for film images on the Hamburg HafenCity development project and a city marketing publication with a 'Dockland' office building (see figs 10.5 and 10.6).

Adam Lampton for the photo of a mock hotel room at the Casino Training School, Taipa, Macao, 2007 (see fig. 11.2).

List of Contributors

Ludger Basten is Professor of Economic and Social Geography at the Department of Sociology at the Technical University in Dortmund, Germany. His research focuses on issues of economic, social and spatial development as well as on planning and governance, in particular with respect to urban and metropolitan spaces and regions in Germany and North America. He has written *Postmoderner Urbanismus: Gestaltung in der städtischen Peripherie* [*Postmodern Urbanism: Design in the Postmodern Periphery*] which draws on empirical work undertaken in Germany and Canada. More recently he has edited a bilingual volume entitled *Metropolregionen – Restrukturierung und Governance. Deutsche und internationale Fallstudien* [*Metropolitan Regions – Restructuring and Governance. German and International Case Studies*] published in 2009.

Sybille Bauriedl is Assistant Professor at the Department of Political Sciences at the University of Kassel, Germany, and associate fellow of a research project on environmental governance and regional climate change. She was research fellow for several projects concerning urban and regional development of the European metropolis and lecturer for feminist geography at the Departments of Geography at the Universities of Hamburg and Munich. Sybille has published several articles in the research fields of sustainable development, urban politics, feminist/queer geography and discourse analysis. In 2007 she published a book on power structures of urban development discourses in Hamburg (*Spielräume nachhaltiger Entwicklung*), in 2008 co-edited a book on urban futures of European cities (*Stadtzukünfte denken*) and she has just published a book on the co-constitution of gender relations and space (*Geschlechterbeziehungen, Raumstrukturen, Ortsbeziehungen*).

Michael Dear is Professor of City and Regional Planning at the University of California, Berkeley, and Honorary Professor in the Bartlett School of Planning at University College, London (England). Michael's current research focuses on comparative urbanism, and the future of the US–Mexico borderlands. Michael was founding editor of the scholarly journal *Society and Space: Environment & Planning D,* and is a leading exponent of the Los Angeles School of Urbanism. His books include: *From Chicago to LA: Making Sense of Urban Theory, Postborder City: Cultural Spaces of Bajalta California,* and *The Postmodern Urban Condition,* which was chosen by CHOICE magazine as an 'Outstanding Academic Title' in 2000. His latest edited volume, entitled *Geohumanties: Art, History, Text at the Edge of Place,* will be published in 2010. Michael has been a Guggenheim Fellowship holder, a Fellow at the Center for Advanced Study in the Behavioral

Sciences at Stanford, and Fellow at the Rockefeller Center in Bellagio, Italy. He has received the highest honours for creativity and excellence in research from the Association of American Geographers, and numerous undergraduate teaching and graduate mentorship awards.

Ulrike Gerhard is Associate Professor of Urban Geography at Würzburg University, Germany. For the past several years, she has specialized on globalization and urban development, looking at economic, political, and social aspects of North American and European cities. Among her publications is a book on the political urban geography of Washington, D.C. (*Global City Washington, D.C. Eine politische Stadtgeographie* 2007), on consumer behavior in Canadian cities (*Erlebnis–Shopping oder Versorgungseinkauf? Eine Untersuchung über den Zusammenhang von Freizeit und Einzelhandel am Beispiel der Stadt Edmonton, Kanada* 1998), as well as several articles on urban design and planning, urban discourses and retail developments in cities.

Jürgen Hasse is Professor of Geography and Geographical Education at the University of Frankfurt, Germany. His fields of reseach have been the relationship between humans and nature, spatial sozialization, urban aesthetics and phenomenological studies. Among his numerous publications are *V. Jahrbuch für Lebensphilosophie: Gelebter, erfahrener und erinnerter Raum* (2010/2011, with R. Kozljanič). *Unbedachtes Wohnen: Lebensformen an verdeckten Rändern der Gesellschaft* (2009), *Die Stadt als Wohnraum* (2008), *Übersehene Räume* (2007), *Fundsachen der Sinne* (2005), *Graf Karlfried von Dürckheim: Untersuchungen zum gelebten Raum* (2005), and *Die Wunden der Stadt* (2000). Jürgen Hasse is a member of the executive board of the German Gesellschaft für Neue Phänomenologie [Society of New Phenomenology].

Jacques Lévy is a Geographer and Professor at the Swiss Federal Institute of Technology, Lausanne (EPFL). He is the director of the Chôros Laboratory and co-director of the Collège des Humanités. Lévy has been a visiting professor in various universities including UCLA, NYU, USP (São Paulo), L'Orientale (Naples), Macquarie (Sydney), Wissenschaftskolleg (Berlin) and he has held the Reclus Chair in Mexico City. His major concerns are the social theory of space, urbanity, globalization, cartography, and the epistemology of social sciences. Among his numerous publications, the following are of particular note: *Géographies du politique* (1991), *Le Monde: Espaces et Systèmes* (1992/1993, with M.–F. Durand and D. Retaillé), *Egogéographies* (1995), *Le Monde pour Cité* (1996), *Le Tournant Géographique* (1999), *Logiques de l'Espace, Esprit des Lieux* (2000, with M. Lussault), *From Geopolitics to Global Politics* (2001, with F. Cass), *Dictionnaire de la Géographie et de l'Espace des Sociétés* (2003, with M. Lussault), *The City* (2008), and *Our Inhabited Space* (2009).

Achim Prossek is Assistant Professor at the Faculty of Spatial Planning at the Technical University in Dortmund, Germany. His current research focuses on visual politics, representations and intercultural aspects of the Cultural Capital of Europe, Ruhr.2010. His latest publications are *Bild-Raum Ruhrgebiet. Zur symbolischen Konstruktion der Region* (2009) and *The European House: Museum and Supermarket* (2009), in *China and Europe. The Implications of the Rise of China for European Space*, edited by K.R. Kunzmann, W. Schmid and M. Koll-Schretzenmayr. London, New York, 209–18. Achim is also co-editor of the *Atlas der Metropole Ruhr* (2009).

Wolf-Dietrich Sahr is Professor of Social and Cultural Geography at the Federal University of Paraná, Curitiba, Brazil, and also lectures sporadically at the Universities of Heidelberg and Karlsruhe, Germany. In 2003, he has been a Visiting Professor at the Centre for Intercultural Studies at the University of Mainz, Germany. His current research focuses on aesthetics and geography, with special reference to expressions of rural and urban spaces, intercultural studies and geography of religion. Most of his recent epistemological work is based on theoretical approaches beyond post-structural and representational philosophies. He has published in a number of books and journals in different European and American countries, including Germany, Brazil, Poland, France, Great Britain, USA, Netherlands, Czech Republic, Cuba, Colombia, and Jamaica. He has been a co-founder of the regular meetings of *Neue Kulturgeographie* [*New Cultural Geography*] in Germany (2003), and the *Núcleo de Estudos sobre Espaço e Representações* (NEER) in Brazil (2004), both well-known networks that are involved in the development of Postmodern Cultural Geography in these countries.

Heiko Schmid is Visiting Professor of Geography at the University of Jena, Germany. He specializes in urban and cultural geography and focuses on themed urban environments, the commodification of lived spaces, and aspects of attention and fascination. Within a phenomenological and action theory approach Heiko developed a research perspective called 'Economy of Fascination' providing one of several inspirations for the current book. Among his publications are a monograph on the reconstruction of downtown Beirut, *Der Wiederaufbau des Beiruter Stadtzentrums* [*The Reconstruction of Downtown Beirut*] (2002), and a recently published book on *Economy of Fascination: Dubai and Las Vegas as Themed Urban Landscapes* (2009). He was awarded the *Hans-Bobek Preis* of the Austrian Geographical Society in 2003 and the *Wissenschaftspreis für Anthropogeographie* of the German Frithjof Voss Foundation in 2009.

Tim Simpson is Associate Professor and Associate Dean of the Faculty of Social Sciences and Humanities at the University of Macau, where he teaches courses in the Communication Department, Gaming Management Program, and Honor's College. His current research interests are in the subjects, spaces, and practices of post-socialist consumption; and in the political economy of intercultural relations.

Neil Smith is Distinguished Professor of Anthropology and Geography at the City University of New York, where he founded the Center for Place Culture and Politics, and also the Sixth Century Chair of Geography and Social Theory at the University of Aberdeen. His research explores the broad intersection between space, nature, social theory and history. Neil's environmental work is largely theoretical, focusing on questions of the production of nature. His urban interests include long term research on gentrification focusing especially on the 'revanchist city' which has filled the vacuum left in the wake of liberal urban theory. He has published various books including *The Politics of Public Space* (2006, with S. Low), *The Endgame of Globalization* (2005), *American Empire: Roosevelt's Geographer and the Prelude to Globalization* (2002), *The New Urban Frontier: Gentrification and the Revanchist City* (1996) and *Uneven Development: Nature, Capital and the Production of Space* (1984/1991/2008). Neil's work has been translated into many languages, he sits on numerous editorial boards, and is an organizer of the International Critical Geography group.

Anke Strüver is Professor of Social and Economic Geography at the University of Hamburg, Germany. She received a PhD at the Department of Human Geography, University of Nijmegen, Netherlands, and her research focuses on the co-constitution of identities and spaces on various scales. Her books include *Stories of the 'Boring Border': The Dutch–German Borderscape in People's Minds* (2005) and *Macht Körper Wissen Raum? Ansätze für eine Geographie der Differenzen* (2005); her most recent edited volume is entitled *Geschlechterverhältnisse, Raumstrukturen, Ortsbeziehungen: Erkundungen von Vielfalt und Differenz im Spatial Turn* (2010, with S. Bauriedl and M. Schier).

John Urry is Distinguished Professor of Sociology and Director of the Centre for Mobilities Research at Lancaster University, Great Britain. His areas of research have been the relationship between society and space, consumer and tourist-related studies, complexity theory, and mobilities research. He has published various books including *Economies of Signs and Space* (1994, with S. Lash), *Consuming Places* (1995), *Contested Natures* (1998, with P. Macnaghten), *Sociology Beyond Societies* (2000), *Bodies of Nature* (2001, with P. Macnaghten), *The Tourist Gaze* (1990/2002), *Global Complexity* (2003), *Performing Tourist Places* (2004, with J.O. Bærenholdt, M. Haldrup and J. Larsen), *Mobilities, Networks, Geographies* (2006, with K. Axhausen and J. Larsen), *Mobilities* (2007), *After the Car* (2009, with K. Dennis), and *Mobile Lives* (2010, with A. Elliott). He is one of the founding editors of the journal *Mobilities*, and has been the editor of the *International Library of Sociology* since 1990. John Urry chaired the UK's Research Assessment Panel in Sociology in 1996 and 2001.

Ingo H. Warnke holds the Chair of German Language and Interdisciplinary Linguistics at the University of Bremen, Germany; he was former Professor of Sociolinguistics at the Center for the Study of Language and Society at the University (CSLS) of Bern, Switzerland. His current research focuses on a language-based analysis of urban spaces. He is founder and coordinator of the Urban Space Research Network (USRN) which is interested in urban space not only as a developed surface but also in view of communicative action as a dimension of urban constructional process. Among other distinctions Ingo was Research Fellow at Harvard University and the German Historical Institut (GHI) in Washington, D.C. and he has held several visiting professorships in Europe. He has published numerous books and papers on various topics of linguistics, in the recent past mainly about communication in urban spaces.

Chapter 1

Cities and Fascination:
Beyond the Surplus of Meaning

Heiko Schmid, Wolf-Dietrich Sahr, John Urry

In the course of economic and cultural globalization most large cities and metropolises have undergone far-reaching transformations induced by an increasing commercialization of urban space and, as a 'postmodern' consequence, intense processes of semiotization and emotionalization. Thus, urban configurations are more and more supplemented by semiotic and psychological processes. When hypermarkets and shopping centres turn into themed environments, when pedestrian zones become festival market places, when traditional housing is transformed into gated communities, and when even lower-class housing areas are increasingly covered with cultural expressions and advertisements, then the urban environment appears much more semiotic than traditional approaches of sociological and geographical research can capture. The induced social modifications result in new power relations, where rational competences and responsibilities are increasingly transferred to socio-psychological and cognitive processes. This new social configuration is accompanied by new constellations of urban actors, diminishing the influence of public actors and transferring their power to private and semi-state actors.

Such a situation calls for a more comprehensive approach combining semiotic, cognitive and emotional elements. The growth of private urbanity through spectacularization is followed by the creation of new 'urban' subjectivities. These are partly based on consumer attitudes that refer to hedonistic and mass-oriented behaviour. Several classic authors have already pointed to this phenomenon. Initially, they referred to the 'bourgeois' as the new rising class, as predicted poetically by Baudelaire and the symbolist French poets. Some decades later, Thorstein Veblen's famous *The Theory of the Leisure Class* (1899), and Georg Simmel's writings about the emotional consequences of life in large cities, as well as the profound reflections of Walter Benjamin on the passages of Paris, have put such a new socio-psychological disposition in scientific terms. But it was only when Guy Debord published his famous *La Societé du Spectacle* [*Society of the Spectacle*] (1967) that the phenomenon of theming became recognized as a socio-political structuration element in the city, revealing the function of the aesthetic in forming urban space.

Consequently, since the 1980s and 1990s, several authors have increasingly focussed on the semiotic effects of this phenomenon, principally in the French and

English speaking communities. In 1986, Gottdiener and Lagopoulos in *The City and the Sign* related signification to material culture. Later on, Mark Gottdiener (1995) examined aesthetical processes based on a specific socio-semiotic approach, examining examples such as shopping malls. Other authors, like Lash and Urry (1994) in *Economies of Signs and Space*, addressed the phenomenon of a commodification of signs from the consumption side, interpreting the aestheticization of material objects as crucial in an emerging economy of signs. Among German speaking authors, we highlight a slightly different discussion, with Georg Franck's approach of an *Ökonomie der Aufmerksamkeit* [*Economy of Attention*] (1998), Gerhard Schulze's reflections of *Die Erlebnisgesellschaft* [*The Event Society*] (1992), Bernhard Waldenfels' *Sinnesschwellen* [*Senses' Thresholds*] (1999) or Hartle's *Der geöffnete Raum. Zur Politik der ästhetischen Form* [*The Opened Space – Politics of the Aesthetic Form*] (2006). While most English speaking authors have referred to critical and/or semiotic, including neo-Marxist, approaches, German speaking authors usually refer to more phenomenological perspectives. However, rarely have both tendencies been discussed together. Therefore this book tries to open up such a dialogue, in spite of some aspects of incompatibility, between these different approaches.

Specifically, Georg Franck's contribution, an *Economy of Attention* (1998), tries to bridge the diverse perspectives. Franck analyzes processes of scarcity, as a basic principle for price mechanism in capitalism, but also as a socio-psychological process of individuation that focuses on attention. Attention then is interpreted as a specific form of 'income', gaining its value as reputation or prominence, as 'payment of attention', mingling economic, psychological, and phenomenological factors. In this perspective, the actual city appears as a battle field for attention, and not only as an expression of use and exchange value as it was commonly marked in classical economic approaches. Psychology and phenomenology now assumed the function of a kind of parallel economy. In this respect, fascination – as a component of socio-psychological attention – is embedded into a wide ranging field of emotional attitudes, like fear, rejection, and repulsion on the negative side, and relaxation, comfort, and attraction on the positive side. What is at stake here, is on one hand the psychological forming of an urban sensitivity and cultural emotional competence, but also the manipulation of attraction and repulsion by urban actors. Thus, psychological attitudes and stage-managed events often have a magnetizing and captivating effect on the consumer, so that economic and social decisions are no longer rational. Perception and consumption of experience becomes a 'vice'. In this sense, 'fascination' is not only described as an externally structured compulsion and binding process (enchantment), but also forms the internal desire for seduction and happiness. It thus plays a key role in the commercialization of theming and in the economic utilization of urban landscapes. An 'Economy of Fascination' (Schmid 2006, 2009) is situated right in this intermediate field between economy, psychology, and phenomenology. It becomes an effective link between economic rationality and emotionality in the life world, which requires an intense and dialogical reflection of both critical and phenomenological attitudes. Therefore, this book is an attempt to bring together these sometimes very diverse theoretical and/or empirical aspects that highlight the interface between ratio and feeling by focussing on the fact of fascination in the city.

Fascination from Enchantment to Delight: (Etymological) Roots as Representations of Social Transformations

The word fascination has passed through several reinterpretations and resemanticizations, changing from a word meaning witchcraft to one associated with social experience and popular culture (Weyand 2009: 208). This transformation has not been linear, but has taken place in several shifts, moving from magic to interpersonal relations, and then to the attraction for things and events in general: 'The history of the term goes back to antiquity, linked for a long time with the belief in witches until it developed in the nineteenth century to a form of interpersonal gravitation and since the twentieth century to an attractiveness of things and events of all types.' (Hahnemann and Weyand 2009: 9).[1]

The term fascination originates from the Latin *fascinus*, which means spell or witchcraft. Among linguists, it is debated if *fascinus* itself has been borrowed from the Greek *báskanos* [sorcerer, bewitcher] or if *báskanos* only stands for 'to speak' traced back to Thracian *bask* (Klein 2003: 274, Barnhard 2002: 370). In the centuries following antiquity, 'fascination' in Europe was still related to the meaning of witchcraft, although the word was not excessively used in the Middle Ages. It was the witch hunting era in the early modern period, between the late fifteenth and early nineteenth century, that induced a certain (re)naissance of the term. In European literature, 'fascination' is documented for the first time in the late sixteenth and early seventeenth centuries: e.g. in Francis Bacon's *Advancement of Learning* from 1605. Symptomatic for the sporadic use of the terminus at that time is William Shakespeare's tragedy *Macbeth*, in which the author addresses the topic of witchcraft, prophecy, and suggestion, but without ever using the word 'fascination' (Seeber 2009).

In the era of Enlightenment 'fascination' gradually veers away from the semantic field of witchcraft becoming more and more associated with interpersonal attraction. Several patterns of explanation exist for this process of resemanticization: one explanation is the rationale-based rejection of any mystic idea of witchcraft. At that time the acceptance of a 'catching' eye (contact) was interpreted by some authors as a 'visual' infection, as it was by the German physician Sennert.

---

1 'Die Geschichte des Begriffs reicht zurück bis in die Antike, verbindet sich lange Zeit mit dem Hexenglauben, bis sich die Faszination im Laufe des 19. Jahrhundert zu einer Form der zwischenmenschlichen Anziehung und ab dem 20. Jahrhundert zur Anziehungskraft von Dingen und Ereignissen jeglicher Art entwickelt.'

A more socio-psychological perspective was introduced through the principle of sympathy. Following this idea, the German philosopher Sloterdijk (1998: 212) interprets fascination as sympathy that interpersonally originates from the reciprocity of love, a romantic bourgeois concept that arises in the seventeenth century and matures in the Enlightenment. In this context erotic aspects and the (corporeal) desire are addressed, so that Sloterdijk (1998: 261) speaks of fascination as 'a magical state of being enravished in mutually erotic enchantment.'[2] One prime expression of this feeling can be found in Baudelaire's lyrics of the nineteenth century, when he speaks of fascination as an almost mesmerizing encounter with a beautiful woman, combining an aesthetic situation of unexpectedness, spirituality, and erotic desire: 'The sweetness that fascinates and the desire that kills' (Baudelaire, quoted in Seeber 2009: 97).[3] The word 'mesmerize', very common in French spiritualism as a counter reaction to rationalist positivism, leads to a specific physical aspect in the interpersonal social relations of fascination: that of hypnosis, the manipulation of consciousness. In late eighteenth century Europe, the 'disease-philosophical approach'[4] of the German physician and astrologist Franz Anton Mesmer had become very popular (Sloterdijk 1998: 228). Mesmer described the influence of the planets on the human body and on disease and used this idea for the treatment of patients through hypnosis: 'The power of the mesmerist to fascinate or entrance his subjects was most commonly explained as the effect of magnetic or electrical forces originating in the body of the mesmerizer and passing across to his subjects.' (Connor 1998: 11). Although the idea of mesmerism was heavily disputed, it had established itself as an almost scientific theory for interpersonal relations.

Sloterdijk (1998: 257ff.) also points to a third tradition of fascination, mentioning Johann Gottlieb Fichte's *Ekstase der selbstlosen Aufmerksamkeit* [ecstasy of a selfless attention]. The German philosopher of Enlightenment had adapted the idea of Mesmer's magnetism to his idealism and assigned it to the situation of a student paying his full attention to the rhetoric of his teacher. The student's attentional ecstasy in this situation symbolizes, according to Fichte (1835, quoted in Sloterdijk 1998: 260f.), a devotion and self-abandonment before God, reflecting the *Grundpunkt der Individualität* [point of origin of individuality].

In the twentieth century 'fascination' developed a more general meaning in the sense of delight and attraction. This conversion into a psychological power coincides with a shift from a transitive to an intransitive relation, and therefore points to an indefinite and vague form of attraction. Now, fascination is no longer attributed to magic and interpersonal attraction: 'Attention has shifted away from the power to fascinate possessed by certain persons or beings, to the power of fascination as it may be lent to or bestowed upon fascinating objects, ideas or persons.' (Connor 1998: 12). Leaving most of its former semantic ballast behind, it became a widely adapted term of popular culture. But at the same time,

---

2   'magische Hingerissenheit im erotischen Gegenseitigkeitszauber.'
3   'La douceur qui fascine et le plaisir qui tue.'
4   'krankheitsphilosophischer Ansatz.'

'fascination' retained a kind of mystification by its inherent vagueness and by not revealing its own secret completely (Seeber 2009). This makes 'fascination' a variable 'projection screen for the imagination of the consumer'[5] as Hahnemann and Weyand proclaim (2009: 7). Drawing on Roland Barthes's theorization on photography in *La Chambre Claire [Camera Lucida]* (1980), Hahnemann and Weyand argue that 'fascination' is an aesthetic experience that is based on an inner restlessness that occurs when something can not be denominated (2009: 22).

Against this background, it seems that 'fascination' is construed as a force that includes attraction, desire, and mystification, but also terror and fear, far beyond a materialism and conception of the rational subject.

Ambivalences as Definite Characteristics: Gazes and Dualities

Departing from this etymological background of 'fascination', which represents the transformative history of its social function, this book develops theoretical and empirical ideas for fascination that focus on the phenomenological aspects of proximity and distance, the psychological permutation of self and not-self, the ambivalence of positive and negative aesthetics, and especially the commercialization of fascination that intrudes upon and emotionally shapes the consumer's lived experience.

The dichotomy of proximity and distance is probably the most important aspect in the phenomenology of fascination. To a certain extent, fascination neutralizes the Euclidean distance by a 'touching gaze'. This is particularly evident in the writing of French novelist Maurice Blanchot, who situates, as Weingart (2002: 24) expresses it, 'fascination in the tension of distance and touch, of separation and encounter, of blindness and vision'.[6] Fascination overcomes, and at the same time results from these dichotomies: 'Whoever is fascinated doesn't see, properly speaking, what he sees. Rather, it touches him in an immediate proximity; it seizes and ceaselessly draws him close, even though it leaves him absolutely at a distance.' (Blanchot 1989: 33).

The lack of distance in perception and the process of distancing between the observer and the observed has been conceptualized by the German philosopher Bernard Waldenfels in his *Phänomenologie der Aufmerksamkeit [Phenomenology of Attention]* (2004). Waldenfels highlights a *Doppelbewegung* [reciprocal dynamic] between salience and attending, between the 'things' that are salient and the 'subjects' who attend to things. He thus proposes an intermediary between 'things' and 'subjects' that can be represented through various techniques and social practices that incorporate power and control in politics and an economy of attention (Waldenfels 2004: 228).

---

5   'Projektionsfläche für die Imagination der Konsumenten.'
6   'Faszination in der Spannung von Distanz und Berührung, von Trennung und Begegnung, von Blindheit und Vision situiert.'

The dissolution of distance between the self-determined subject that is a constructive principle of modernity, and the objects of his (also constructed) desires in a postmodern capitalist system induce new aesthetic-moral attitudes, both in an affirmative and a rejective stance. The affirmative includes a gaze, where imagination leads to the satisfaction of a supposed inner desire. The negative aspect is related to the sphere of manipulation, where mechanisms of infatuation and mesmerization consciously or unconsciously distort the free will of a person. This gives 'fascination' a social and political ambivalence that, following Connor (1998: 12) appears as a kind of self-determined arrest: 'The desire for fascination is a desire for arrest, but of a certain enlivening kind, in which the subject of fascination is at once enthralled and aroused.' Thus, the modern subject is caught between their own desires as supposed self-expressions and their consciousness about manipulation processes that might have caused these through seduction.

Desire and seduction are also key processes in the commodification of this phenomenon. Here, fascination serves as an emotional multiplicator in the consumption process. It addresses emotional needs, while offering satisfaction. Due to the fact that such satisfaction is limited to a short moment, a continuous process of desiring is induced calling for new satisfaction – sometimes in an addictive way. Against this background, fascination is most attractive for an expanding market economy. Industry now produces desires and addictions, using processes such as excessive attention and stimulus satiation, and thus relies basically on human emotions.

Fascination represents a socio-psychological construction of a specific (incomplete) duality. It cannot be forced by means of enchantment alone, but is always dependent on the disposition to 'receive an external will' (Weingart 2009: 48).[7] Schmid (2009: 25) has expressed this relation between economy and emotion as follows: 'Fascination works in two respects: on the one hand it is an externally structured compulsion (enchantment), and at the same time it takes the form of an internal desire on the part of the consumer, which is directed at the exotic and the unique, and also at the thrill of gratification and seduction.' From this perspective, fascination is based on mainly individual (sometimes also collective) desires and political structuration. In a capitalist economy, it appears in the space of 'inbetweenity' of producing a psychologically construed self (the consumer) and a gigantic system of structures of seduction (the market), where even profit (understood as the transformation of the Nietzschean desire for power into a market equivalent) turns out to be a secondary element. Such 'inbetweenity' of fascination explains, why epistemologically the authors of this book argue in such a diverse manner – sometimes they try to understand the phenomenon from a more theoretical position, torn between phenomenological, neo-Marxist, and/or (post-) structuralist approaches, sometimes they turn to concrete examples, expressions, forms and structures of fascination in the city, and some – even going further – highlight the desired and undesired consequences of such a social transformation.

---

7   'einen fremden Willen zu empfangen.'

In all the contributions, however, it is revealed how in late modern capitalism the spheres of psychology, society, and ecology have been re-ordered or disordered to something new.

Fascination and urbanity: a complex dimension and dimensions of complexity

From this analytical perspective, 'fascination' in the city is not a simple phenomenon, but addresses a complex set of characteristics. We distinguish four pivotal dimensions which are all tackled in this book: (1) aesthetics, (2) emotions, (3) lived experience, and (4) power structures and governance.

As the urban environment is a built environment, formal physical changes under the auspices of *aesthetics* are of prime importance. Aesthetics here does not simply represent the question of beauty and cleanliness, but the formal construction of a space as social means. In this context, fascination is used as a stimulus (and agitation), which exceeds mere functional aspects and values, for the restructuring of the symbolic dimension of identity, status, and power through the creation of focal points in the urban landscape.

The *emotional* context presents fascination as a psychological link that incorporates fetishism, preference, but also dislike, rejection, or fear. Here, fascination establishes a mostly non-reflexive psychological nexus between the inhabitants of a city and their environment on the edge, between ratio and emotion.

Closely linked to the emotional context is the question of *lived experience*. Here, fascination represents a set of cognitive or performative acts that conquer and/or lose, or at least occupy, portions of the urban space as an environment for survival, self-fulfilment and self-expression. The lived experience of fascination has acquired different aspects in the city; these aspects range from developing cognitive capacities in lived space to the bodily felt embeddedness in the urban environment.

Finally, within networks of power relations, fascination incorporates the three aforementioned factors by conditioning social relations through new *power structures and forms of governance*. In fact, nowadays fascination, whether it includes fear, or a positive binding, has become a powerful means of governance in our daily life experiences. Therefore, power relations are central to the question of public space as an aesthetic and emotional landscape, to the problematic use of commercial space as a semi-public space, and to the re-ordering of life in urban lived spaces in the sense of who orders what. In this regard, power is omnipresent and inherent to fascination not only in its magical capacity, but also as a rational form of urban structuring and binding.

As can be seen, the socio-psychological processes of fascination are complex and differentiated in urban space. They reveal the horizon for a concept of a new urban 'self', a 'self' that reproduces – as is common in postmodern times – its own fragmented configuration via the spatial frictions of postmodern cities, divided between the public and the private spheres, distributed between commercial and cultural places, surviving in a mosaic of heterogeneous lived spaces, and embedded into a network of ordered socio-political spaces.

Paths to Fascination

All the contributions of this book expand on one or more of the dimensions mentioned above to develop their analysis. The different perspectives and standpoints of the authors are divided between contributions that are of more theoretical (Chapters 2 to 6, and 12), and others that are more empirical (Chapters 7 to 11), though all are tied via a sometimes more hidden, sometimes more overt discussion on the different aspects mentioned above.

Michael Dear reads postmodern urbanism and the role of fascination against the background of aesthetics and lived experiences, venturing into downtown cores and suburban privatopias. He refutes the actual debate on urban theming as a merely economic and political attempt to erase individual experiences, and reminds the reader that social structuration is profoundly dependent upon cognitive and emotional factors. Therefore, while exploring the urban reality, Dear refers to the modern/postmodern divide as an epochal and epistemological change, which recontextualizes capitalism and subjectivity. For this purpose, he proposes the concept of Keno capitalism as a capitalism that is based on the individualization of the urban experience. Keno capitalism, in this sense, leads to the hollowing out of urban governments and causes the fragmentation of urban space into new socio-spatial autonomies. As a result, Dear observes a radical aesthetic flattening, a pastiche of homogenous brands in a desert, where some spots are marked by spectacular urbanism that punctuate the urban landscape through fascination, while others are featured by incontrovertible ugliness. Thus, modern homogeneity has been increasingly lost to individualized cognitive worlds, which again are structured through an all-encompassing capitalism.

While Dear refers to the object of fascination in the city, *Jacques Lévy* critically reflects the researchers' positions on the phenomenon of the city itself. He confronts 'urban' theoretical approaches with the bodily lived possibilities of modern urbanity. In a kind of negative archaeology, Lévy uncovers the rural roots in most urban theories, using as examples both North American urban sociology and Heidegger's ontological ruralism. Opposing such a theoretical 'city of the angels' to the city of bodies, he unveils the urban experience as co-presence (understood as co-operation and competition), serendipity, and the infinitude of the humane process of urbanization. In this context, he highlights the simultaneity of the 'real' and 'representation' as a basic foundation of lived urbanity. Lévy proposes from this perpective that the 'fascination issue' resides more in urban studies than in the life world of urban dwellers themselves.

Jürgen Hasse investigates the postmodern aesthetization of urban environments. His stance is based on the discussion of feelings and emotions of urban dwellers as specific forms of urbanity. Venturing into the 'logics' of settings beyond the logics of 'meanings', his reflections are directed towards the divide between presentation and representation. His phenomenological concept is based on the idea of 'surroundings', as proposed by Count von Dürckheim at the beginning of the twentieth century, and the current discussion on 'atmospheres' through members of the so-called *Neue Phänomenologie* [New Phenomenology] in Germany (Böhme, Schmitz) as forms of presentation. Thus, Hasse goes far beyond rationality and representation in a conventional sense, and identifies aesthetic expressions as appealing to both individual and collective feelings and emotions as a form of 'living in the city'. Using the examples of urban illumination, urban garden landscapes, and symbolic architecture, Hasse leads the reader to the question of if and how emotional modes, such as fascination, can be scientifically captured via rationalistic approaches.

Emotions are closely linked to perceptions, and so *Ludger Basten* points the debate of New Urbanism in this direction. Highlighting methodologically the crucial importance of the comprehension of postmodern urban transformations, he investigates the divide between perceptions from the outside in contrast to perspectives from within. Thus he demonstrates that the creation of peripheries is based on the ideological production of images of urban developers who transform their ideas into built environment, but also insists that the peripheral dwellers themselves take their chance in remodelling their areas with their specific aesthetics in ordinary everyday life, as telling images of their own urbanity. So as in the centre, the periphery is marked by fascination as an act of immediacy. But while outsiders sell 'fascination' (marketing) as an immediate impression (trying to avoid critical reflection), the urban dwellers themselves develop a different type of fascination, this time as a space of possibilities where the flattened pastiche of images can be deepened, reuniting appearance and reality in authenticity.

Neil Smith enters into these debates from a Marxist perspective, drawing on Marcuse and Lefebvre to focus on the important difference between appearance and reality. Going against poststructuralism and the empiricism of English-speaking cultural geography (including its idealistic implications), he argues that the contemporary debate on fascination mixes both categories as a 'knack' and thus risks fetishizing the landscape of the built environment rather than analyzing the social processes that create and inspire the 'tango' of appearance and reality. Referring to Lefebvre, Smith shows that any creation of a material space has an immediate effect on the social structuration of urban society, on urban experience, and on social struggle. Smith's argument passes through an examination of two issues. First, he points to the realities of what he calls 'urbicide', which can be encountered in the recent intensification of urban warfare in techno-military landscapes, exercised in capitalist state actions; here, fascination is evoked as a manipulated means of social action. Second, he widens the meaning of the state when going beyond traditional state functions like planning and control, understanding fascination (like Basten in this book) as an ambiguous form of urban experience, which fully involves individual experiences and also governmental intentions. In this context, Smith identifies fascination as a relation to the urban environment that is positioned in what he calls a 'crawl space' between appearance and reality, duping on one hand by the effects of superficiality, but on the other appearing as a resource for deepening reality beyond a rational approach.

Wolf-Dietrich Sahr for his part focuses on the question of 'reality' in a different way. Referring to the importance of value spheres, commonly seen as subjective 'realities', he investigates the dialectical game of urban planning strategies and urban experience by using the example of the Brazilian 'model' city of Curitiba. Avoiding the traditional bipolarity of use and exchange values as relational values, Sahr points to the relevance of absolute values in the subjective urban experience. Drawing on Guy Debord's reflections on spontaneity and Deleuze's and Guattari's concept of 'face value', he interprets the urban landscape as a locale of 'inbetweenity' between concept and experience. In his example, he demonstrates that the creation of fascination through the construction of urban projects, which in Latin American cities are mostly formed by elite aesthetics, are a mixture of relational signification processes and absolute comprehensions of subjectivation, so that they also can be interpreted as a field of possibilities for the reappropriation of urban space through the city's inhabitants. This is possible when fascination is perceived as an immediate (non-rational) relation with the urban environment that helps to reinforce the autonomy and subjectivity of self-expression.

The dialectics between urban planning and urban experience is also the subject of *Ulrike Gerhard's* and *Ingo H. Warnke's* contribution. The authors draw attention to the question of how fascination can be eroded during the urban planning process, its implementation and the dwellers' reactions. In a careful discourse analysis, they investigate the modernist and participatory urban planning project of Emmertsgrund in Heidelberg, which in the 1960s was highly praised as a progressive planning strategy, profoundly influenced by the discourse of the Frankfurt School of Critical Theory. In their reflections, the authors demonstrate that the 'ideal city' in its utopian function is not fixed, but shifts around and passes through different meanings under the changing influences of political interests and socio-economic conditions. Thus, the discourse of an urban project is permanently intertwined with social experience. Here, the horizontal elements of the discourse, as formulated in a plan and some intellectual reflections, are gradually deepened during the implementation phase of an urban project, so that finally the vertical dimension of lived urban space influences the discourse, and the local population becomes its co-author. In this sense, Emmertsgrund shifted its image of fascination from a progressive social dreamland of leftist theorists to what they call a 'wasteland' under the lived experience in a capitalist environment.

In the same vein, but from a more deconstructive stance, *Achim Prossek* describes the attempt of planners to provoke fascination in the Ruhr District, an old, partly run down, region of coal and steel industries in Western Germany. Prossek critically investigates the strategy of moulding the image of a 'European

Capital of Culture 2010' through historic, aesthetic, and touristic imaginations, by characterizing such a process as a 'soft designing' that is embedded into a complex political network of local initiatives, capitalist stakeholders and state agencies. In this context, he carefully analyzes the distorting function of the 'tourist gaze' (Urry) for local identitification processes, when he mentions how even 'locals' are impressed and fascinated by new visions of their home areas through such a gaze. He explains this effect as a result of the atmospheric shaping of a 'metropolis', which only becomes visible at points of vision (hubs) within the urban landscape. Here, tourists from outside and locals respond to the mingling of narrative, image, and social change in a contemplative perspective.

The shaping of images, and the close relation between economic and symbolic processes in postmodern urban planning processes is also the focus of *Sybille Bauriedl's* and *Anke Strüver's* chapter on the Hamburg HafenCity. In their study, the authors investigate the production of urban images and fascination through film industries. Methodically, they try to tie semiotic aspects of image-meanings, especially their representation and interpretation, to discursive aspects, which involve the production, circulation, and adoption of images by urban planners and urban dwellers. Hence, the urban image strategy, which in the case of Hamburg combines traditional harbour elements with the image of a European city and postmodern with its cityscape aesthetics, is revealed as a specific strategic placement that anchors the urban location by virtualization into a broader capitalist space. Thus, the game of place and space uplifts simple 'place values' of the urban experience into 'image values' of fascination, which can be interpreted as a new social and spatial function in this game.

While Bauriedl and Strüver point to the role of visualization in the place-space game of fascination, *Tim Simpson* reports on a corporeal and emotional case of urban fascination. Based on a bridging approach between Harvey's Marxian political economic analysis and Foucault's account of subject formation, he demonstrates in the gambling industry (Baccarat) of Chinese Macau, how the phenomenology of gambling can be interpreted as a neoliberal formation process to construct a specific subjectivity and a specific consumer type (the working consumer) who is embedded – via fascination – into processes of accumulation by dispossession. Thus, in Macau casinos become places where the urban spatial aesthetics of themed environments intertwine with spaces of governance and control under the premises of neoliberal capitalism. In this respect, Macao serves as a laboratory of forced consumption through fascination that complements China's transition into a market economy.

John Urry broadens the horizon of the discussion of 'fascination and the city' by discussing the emergent contradictions that stem from the historical shift of societies of production and discipline to societies of pleasure, seduction, and excess. No longer are economies based on basic needs, at least in the ever expanding centres of the First World, but on produced desires. Thus, Urry analyses the geographic change from a world of specialized and differentiated zones of production and limited consumption to a more mobile, de-differentiated world of sophisticated consumption patterns in a high carbon society. The non-equilibrium system of global capitalism directly threatens the sustainable basis for human survival. This results from the growing forms of mobility and the construction of phantasmagoric entertainment sites (such as Dubai's skyline, see Elliott and Urry 2010, for more detail). Here, capitalism and its child 'fascination' is the only game in town, but it seems to become its own 'gravedigger' as it may come to destroy the global conditions of life upon earth.

The panoramic view of this book on fascination and the city is a first attempt at examining a phenomenon embedded into the ongoing processes of restructuring space and capitalism, society and subjectivity. What is hidden under the layer of meaning, sensitive moulding, and scientific interpretations as a surplus, and what is precariously captured in our epistemologies of aesthetics and form, psychology and emotion, phenomenology and experience, power and control, seems to be an overriding human capacity for autonomous self-creation in the maelstrom of capitalism and desire. However, fascination also allows a social utopian perspective when it is, as Gerhard and Warnke have expressed it in our book, 'anticipating consciousness of the human being rendering possibilities for conceptions that are independent of the concreteness of their realisation.' Thus, following the current shift from collective social structures to individualized experiences, the rarely discussed question of fascination obviously seems to be a prime element in the border zone between urban structures and new subjectivities.

References

Bacon, F. 1605. *Of the Proficience and Advancement of Learning.* London: Tomes.

Barnhard, R.K. (ed.) 2002. *Chambers Dictionary of Etymology.* New York, Edinburgh: Chambers Harrap Publishers.

Barthes, R. 1980. *La Chambre Claire: Note sur la Photographie.* Paris: Gallimard.

Blanchot, M. 1989. *The Space of Literature.* Translated from French by A. Smock. Lincoln, NE: University of Nebraska Press.

Connor, S. 1998. 'Fascination, Skin and the Screen'. *Critical Quarterly*, 40(1), 9–24.

Debord, G. 1967. *La Societé du Spectacle.* Paris: Editions Buchet-Chastel.

Elliott, A. and Urry, J. 2010. *Mobile Lives.* London: Routledge.

Fichte, J.G. 1835. *Johann Gottlieb Fichte's nachgelassene Werke, vol. 3.* System der Sittenlehre. Leipzig: Maner und Müller.

Franck, G. 1998: *Ökonomie der Aufmerksamkeit: Ein Entwurf.* München: Carl Hanser.

Gottdiener, M. 1995. *Postmodern Semiotics. Material Culture an the Forms of Postmodern Life.* Oxford, Cambridge, MA: Blackwell Publishers.

Gottdiener, M. and Lagopoulos, A. (ed.) 1986. *The City and the Sign. An Introduction to Urban Semiotics*. New York: Columbia University Press.

Hahnemann, A. and Weyand, B. 2009. 'Faszination. Zur Anziehungskraft eines Begriffs', in *Faszination: Historische Konjunkturen und heuristische Tragweite eines Begriffs*, edited by A. Hahnemann and B. Weyand. Frankfurt am Main: Peter Lang, 7–32.

Hartle, J.F. 2006. *Der geöffnete Raum. Zur Politik der ästhetischen Form*. München, Paderborn: Fink.

Klein, E. (ed.) 2003. *A Comprehensive Etymological Dictionary of the English Language*. Amsterdam: Elsevier.

Lash, S. and Urry, J. 1994. *Economies of Signs and Space*. London, Thousand Oaks, CA: Sage Publications.

Schmid, H. 2006. 'Economy of Fascination: Dubai and Las Vegas as examples of a thematic production of urban landscapes'. *Erdkunde*, 60(4), 346–361.

——— 2009. *Economy of Fascination: Dubai and Las Vegas as Themed Urban Landscapes*. Urbanization of the Earth, vol. 11. Berlin, Stuttgart: Gebrüder Boerntraeger.

Schulze, G. 1992. *Die Erlebnisgesellschaft: Kultursoziologie der Gegenwart*. Frankfurt am Main, New York: Campus.

Seeber, H.U. 2009. 'Faszination, Suggestion Hypnose und die literarische Kultur Englands vor und um 1900', in *Faszination: Historische Konjunkturen und heuristische Tragweite eines Begriffs*, edited by A. Hahnemann and B. Weyand. Frankfurt am Main: Peter Lang, 91–108.

Sloterdijk, P. 1998. *Sphären, vol. 1. Blasen. Mikrosphärologie*. Frankfurt am Main: Suhrkamp.

Veblen, T. 1899. *The Theory of the Leisure Class – an Economic Study of Institutions*. New York, London: Macmillan.

Waldenfels, B. 1999. *Sinnesschwellen. Studien zur Phänomenologie des Fremden, vol. 3*. Frankfurt am Main: Suhrkamp.

——— 2004. *Phänomenologie der Aufmerksamkeit*. Frankfurt am Main: Suhrkamp.

Weingart, B. 2002. 'Faszinationsanalyse', in *Der Stoff, an dem wir hängen: Faszination und Selektion von Material in den Kulturwissenschaften*, edited by G. Echterhoff and M. Eggers. Würzburg: Königshausen & Neumann, 19–30.

——— 2009. 'Blick zurück. Faszination als "Augenzauber"', in *Faszination: Historische Konjunkturen und heuristische Tragweite eines Begriffs*, edited by A. Hahnemann and B. Weyand. Frankfurt am Main: Peter Lang, 33–48.

Weyand, B. 2009. 'Von Hitler bis zum iPod. Grundzüge einer Kulturpoetik der Faszination im 20. und 21. Jahrhundert', in *Faszination: Historische Konjunkturen und heuristische Tragweite eines Begriffs*, edited by A. Hahnemann and B. Weyand. Frankfurt am Main: Peter Lang, 193–211.

PART I
Revealing Fascination:
Theoretical Horizons

Chapter 2

The Urban Question after Modernity

Michael Dear

> The whole life of those societies in which the modern conditions of production
> prevail presents itself as an immense accumulation of spectacles. All that was
> once directly lived has become mere representation.
>
> (Guy Debord, *The Society of the Spectacle* 1992 [1967], thesis 1).

Consensus has it that we have entered a global urban age in which the majority
of the world's population now lives in cities of one kind or another. Yet there is
precious little understanding about what this trend entails, beyond the customary
Malthusian-inspired fear of mayhem. The proliferation of neologisms describing
emergent urban forms is indicative of intellectual confusion rather than mastery.
They include such descriptors as polycentric, postmodern, patchwork, splintered,
and post-(sub)urban. The places produced by these altered processes are variously
labelled as city-region, micropolitan region, exopolis, edge city, or metroburbia.
The relatively recent accumulation of a variety of large-scale exotic urbanisms
has added to the diversity of nomenclature. For instance, Dubai is heralded as
part of the post-petroleum future (Davidson 2008), or indicative of a global
'economy of fascination' (Schmid 2006). These spectacular urbanisms also
include examples related to natural and human-induced disasters, as demonstrated
by the unforgivable post-Katrina failures in New Orleans, and Michael Sorkin's
contribution on post-9/11 commemoration (Sorkin 2008). The central concern
in my essay is to problematize these recent additions to the landscape within a
broader context of contemporary urbanism.

Guy Debord (1992 [1967]: x) wrote *The Society of the Spectacle* 'with the
deliberate intention of doing harm to spectacular society', which he regarded as
complicit in all that was inauthentic about the urban experience. In his preface
to the book's third edition, he noted that events in the 25 years since the original
publication 'merely confirmed and further extended the theory of spectacle.'
Michael Sorkin's edited collection, *Variations on a Theme Park*, was also published
in 1992; his call-to-arms was motivated by a conspiratorial, even sinister view of
a contemporary city in which the public realm was systematically being erased.
In 2007, two edited collections appeared to confirm the prescient forebodings of
Debord, Sorkin and his contributors. First, in *The Suburbanization of New York*,
Jerilou Hammett and Kingsley Hammett asked if the 'world's greatest city (sic)' is
becoming 'just another town':

Today New York is on its way to becoming a 'theme-park city', where people can get the illusion of the urban experience without the diversity, spontaneity, and unpredictability that have always been its hallmarks. Like the suburbs New York has so long snubbed, the city is becoming more private, more predictable, and more homogenized (Hammett and Hammett 2007: 20).

The second book, *Evil Paradises: Dreamworlds of Neoliberalism* (2007), is a collection of essays assembled by Mike Davis and Daniel Bertrand Monk on the world-wide phenomenon of plutocratic landscapes. Especially noteworthy is the volume's accumulation of diverse global examples, including 'floating city-states ... space tourism, private islands, restored monarchies, and techno-murder at a distance' (Davis and Monk 2007: xiv). Davis and Monk (2007: x) claim that these are evidence of 'the massive, naked application of state power to raise the rate of profit for crony groups, billionaire gangsters, and the rich in general.' The price of paradise, they conclude, is a human catastrophe characteristic of 'terminal, not anticipatory, stages in the history of late modernity' (ibid.: xvi, see also Urry's chapter in this book).

Other analysts are less certain about the diagnosis of urban spectacle, and a more careful reading of the contemporary urban circus is long overdue. The restructuring of oil-rich monarchies such as Dubai is not analogous to the emergence of China as a global industrial giant, no matter how superficially similar their spectacular transformations. Dubai is about the size of Ventura County in Southern California, and reputedly houses one-fifth of the world's construction cranes. In its shift to a post-oil economy, Dubai has plans for a $2.2 billion theme park, the tallest building in the world, a global airport hub, a refrigerated beach and other high-end tourist attractions and facilities. This is very different from an industrializing China, a burgeoning India, or US efforts to revitalize aging urban centres by adding starchitect-designed cultural facilities. We should take care not to be blinded by the bling as we contemplate these new urban landscapes, nor lose sight of ordinary landscapes and the subjective experience of the urban. Steve Pile (2005) reminds us of the cognitive, emotional sides of the urban experience by exploring the imaginative, fantastical, or 'phantasmagoric' aspects of city living, including dreams, magic, vampires and ghosts. No matter how we react to the variety of spectacular urbanisms or evil paradises – most commonly with a mixture of horror and fascination – the lived experiences of people in place is an indispensable component of any calculus of urban inquiry.

From Modernity to Postmodernity

In order to properly contextualize the contemporary urban spectacle, I need to briefly recap the history of the 'urban question'. During the second quarter of the twentieth century, many students of the city began a search for an urban science to explain the modernist city. Two moments were especially emblematic of this effort: one focused on intra-urban structure (i.e., the spaces within cities); the second on inter-urban structure (the spaces between cities); in both cases, spatial structure was regarded as the representational key to understanding the material manifestations of social, economic and political practice. Intra-urban structural analysis gained particular prominence via E.W. Burgess' famous concentric-circle view of urban structure, which became one of the hallmarks of the Chicago School in the 1920s (Burgess 1925). Inter-urban structure was primarily associated with a number of German scholars, such as von Thünen (1966 [1826]), Walter Christaller (1967 [1933]), and August Lösch (1954 [1940]), who drew attention to consistencies in the size and spacing of cities in the landscape, introducing what later became known as 'central place theory'. Based on the precepts of the Chicago School and central place theory, urbanists in the latter half of the twentieth century developed many systematic approaches to describing urban form, process and outcome. For example, urban geographers catalogued a wide range of statistical regularities that often found expression in 'laws' of urban structure both at the intra- and inter-urban scales. Such empirical regularities included the 'rank-size rule' (relating to the primacy of a single major city in the urban hierarchy of each nation), as well as multivariate analysis-derived dimensions characteristic of large cities around the world. The new urban science was felicitously described in the jargon of the times as being concerned with cities as systems within systems of cities. Its version of the 'urban question' invented a scientific method to account for the modernist city.

A distinctly different 'urban question' rose to prominence during the early 1970s when neo-Marxian scholars undertook a critique of the burgeoning urban science. They perceived urban structure as derivative of the broader political economy associated with capitalism and power. Several important works of this era (by Lefebvre 1991, Walker 1981, Harvey 1973, for instance) delved deeply into Marx for a materialist explanation of urban process. In his book *The Urban Question*, Manuel Castells (1977 [1972]) even questioned the epistemological status of the concept of a 'city', characterizing 'the urban' as something epiphenomenal to the more fundamental processes of the capitalist mode of production. However, others insisted on the specificity of the 'urban' as an analytical category (e.g. Dear and Scott 1981), regarding the city as a substantive object worthy of inquiry in its own right. They conceived of the 'city' as a primitive (or fundamental element) in urban theory that was both constitutive of, and a consequence of a socio-spatial process. By the mid-1980s, the 'urban question' had called into question the status of the analytical object called the city in both material and epistemological/ontological terms.

The indeterminacies surrounding the material and mental constructs of urban geography were complicated by a renaissance in social theory during the last quarter of the twentieth century. The rise of post-structuralism, postmodernism, feminism, and postcolonialism, *inter alia*, altered the epistemological and ontological bases of knowing. Broadly speaking, such approaches shifted attention to the cultural realm and questioned the cognitive foundations of truth claims from all persuasions. As a consequence, the pre-eminence of urban science and Marxist science were deeply compromised. By the late-twentieth century Castells (2001), for instance, had shifted to a broader theory of network society that included a more nuanced view of urban geography that included cognitive dimensions such as social identity, semiotics and aesthetics.

At the opening of the twenty-first century, any grounded theory of the city was obliged to confront three tasks: undertaking a fresh reading of emerging urban landscapes and form (including spectacular urbanisms); appropriately contextualizing the urban process in terms of its historical epoch and geographical scale (the problems of periodization and regionalization); and engaging the various ontological and epistemological ways of seeing the cityscape. In what follows, I will use postmodern thought to illuminate these three aspects of the urban question.

At its most basic level, postmodernism drew attention to a set of *distinctive cultural and stylistic practices* that were initially emerged from literature, art and architecture. Fredric Jameson's characterization of postmodernism as the cultural logic of late capitalism set an important precedent, recognizing in the landscape a radical 'flattening', or homogenization, of human experience, as well as a proliferation of aesthetic 'pastiche' in architecture and urban design (Jameson 1991). Since the mid-1980s, when urbanists first began a sustained engagement with postmodern thought, the diversity of emerging urban archetypes has been well-documented. The early work included a focus on theme parks, carceral cities, gated communities, edge cities and the like (Dear and Flusty 1998, Sorkin 1992); today, it is the spectacular urban megastructures that fascinate scholars everywhere (e.g. Davis and Monk 2007, Schmid 2006).

Also during the past two decades, evidence of *epochal change* has been convincingly advanced. Jameson (1991) identified postmodernity as the epoch of late capitalism, drawing attention to the totality of emergent social, economic, political and cultural practices that characterized this stage in capitalism's evolution. Evidence for a significant epochal shift includes the following dynamics: *Globalization*, especially the emergence of relatively few world cities as centres of command and control in a globalizing capitalist world economy; *Network society*, including the rise of the 'cyber-cities' of the Information Age; *Social polarization*, referring to the increasing gap between rich and poor, among nations, different racial groups, genders, and those on either side of the digital divide; *Hybridization*, the fragmentation and reconstruction of identity and cultural life brought about by international and domestic migrations; and *sustainability*, including a widening consciousness of human-induced environmental change. These tendencies come together in the problem of *governance*, to which I will return later in this account. While these five tendencies may find formal equivalence in previous eras (e.g. earlier manifestations of forms of globalization), the present concatenation *is* different because they have never before appeared in concert, never before penetrated so deeply, never before been so geographically extensive, and never before overtaken everyday life at such speed. In short, there has never been anything as globally universal as the rise of the 'Information Age'. It is likely to prove as profoundly transformational as the advent of the Agricultural and Industrial Revolutions of earlier times.

Finally, the enduring importance of postmodernism's *ontological and epistemological challenge* has been widely accepted. The rationalities of the scientific revolution and the Enlightenment have been challenged, replacing a belief in truth, foundationalism and laws with beliefs based in relativism, pragmatism and indeterminacy. The central tenets of modernism have been overridden by a cacophony of different voices, each claiming to offer something that helps understand the world about us. It is not that science has been diminished (although its hegemony has), so much as a realization that its brilliant success in certain applications simply does not extend to every field of human endeavour. Postmodernism was instrumental in reinstating the intuitive and emotional dimensions of knowing and, in so doing, dismissed the idea of a grand theory capable of explaining everything. Under such circumstances, the best we can do is to insist on all ways of seeing, avoid using any theory beyond its range of applicability, and resist attempts to institutionalize any single way of knowing. Philosopher Isaiah Berlin has warned that we simply have to get used to living with the radical incommensurability that exists among various philosophical persuasions (Gray 2006); any tilt, Quixote-style, towards false resolutions can be bought only at the price of a damaging reductionism.

Just as the core beliefs of modernist thought have been displaced by postmodernism's multiple ways of knowing, so has the notion of a universal urban process been dissolved by the multiplying logics that are transforming city-making (Dear 2000: ix). At the core of this transformation is the altered relationship between city and hinterland. In modernist urbanism, the impetus for growth and change proceeded outward from the city's central core to its hinterlands. In postmodern urbanism, this logic is precisely reversed: the evacuated city core no longer dominates its region; instead the hinterlands organize what remains of the centre. By this, I mean that urban space, time and causality are being shifted. Thus, the heterogeneous *spatial logics* that characterize contemporary urban development derive from the 'outside-in', not 'inside-out' as in modernist urbanism; in the sequence of urban development, a centre – if one or more ever emerges – appears *chronologically later* than the peripheries; the direction of *causality* is from periphery to centre, even if (as often happens) this finds expression as an absence of direction. In short, the process of *becoming urban* involves altered structural and functional relationships at the inter- and intra-urban scale that are radically different from those in the modernist city. The multi-centred *city-region* is now constitutive of the 'urban' and the urban question, not the classic mono-centric urban centre.

The distinction between modernist and postmodern urbanisms is at its keenest in the two archetypes: Chicago and Los Angeles (cf. Dear 2002, Scott and Soja 1996). The core-to-hinterland causality of modernist urbanism is best captured in E.W. Burgess' account of the evolution of differential urban social areas in

Chicago. Burgess (1925) and other adherents of the Chicago School observed that as distance from the urban core grew, the city would take the form of a series of concentric rings of diminishing density. Now imagine a city where fragmentation (geographical non-adjacency) and decentredness (polycentrism) are the primary urban drivers: there will be many urban cores, not one; independent edge-cities spring up with allegiance to no one city centre; conventional town centres are no longer at the heart of the urban process; suburbs, understood as peripheral accretions to dominant urban cores, no longer exist; and the agglomeration dynamics that historically produced cities have been replaced. The aggregate of

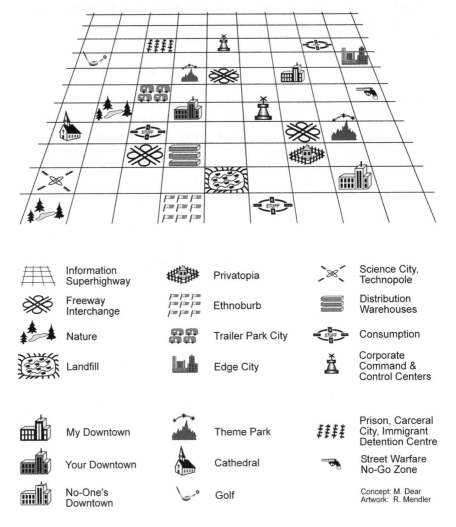

	Information Superhighway		Privatopia		Science City, Technopole
	Freeway Interchange		Ethnoburb		Distribution Warehouses
	Nature		Trailer Park City		Consumption
	Landfill		Edge City		Corporate Command & Control Centers
	My Downtown		Theme Park		Prison, Carceral City, Immigrant Detention Centre
	Your Downtown		Cathedral		Street Warfare No-Go Zone
	No-One's Downtown		Golf		Concept: M. Dear Artwork: R. Mendler

Figure 2.1   A hinterland urban aesthetic

these empirical changes further reinforces the theoretical imperative to question what we understand as a 'city'.

The world of postmodern urbanism is what Steven Flusty and I have called 'keno capitalism' (fig. 2.1). Keno capitalism assumes a world of ubiquitous connectivity courtesy of the Information Age. Urbanization occurs on an undifferentiated grid of opportunities where each land parcel is (in principle) equally available for development as a consequence of access to the information superhighway. Capital settles on a land parcel while ignoring opportunities on adjacent lots, thus sparking urban development. The relationship between development on one lot and another remains a disjointed, unrelated affair, because conventions of urban agglomeration have been replaced by a quasi-random collage of non-contiguous, functionally-independent land parcels. Only after considerable time will these isolated developed parcels collide with others and *take on the appearance* of what we normally regard as a 'city'. However, there is no necessity for such an agglomeration to occur, because the keno capitalism grid is infinitely expandable in any direction, allowing a fragmented urbanism to occur piecemeal for as long as potential development parcels remain available and wired. The notion of centres and edges disappears, as does the distinction between intra-urban and inter-urban process.

By using the term 'quasi-random' I do not mean to suggest that postmodern urban process is *a*-logical, merely that it is characterized by multiple logics that can override (or at least mask) earlier conventions of urban form and process. Nor do I assert that modernist urbanism is dead, although it may everywhere be ailing. Just as some places in the American west and south-west are already postmodern in their urban process, many older cities in the north-east and mid-west retain their residual modernist ways. However, even places of persistent modernism (including Chicago) are now being over-written by the texts of postmodern urban process, providing compelling evidence of urban and social change in the Information Age.

Redefining the Urban Question: Sprawl, Sustainability, Politics

The altered textual, epochal and epistemological/ontological dimensions of postmodern urbanism require a new urban lexicon that re-categorizes urban concepts and (ultimately) redefines the urban question. For instance, the notion of a single dominant 'downtown' core is best understood as an historical expression of modernist urbanism; downtown Los Angeles cannot be construed as *the* nerve centre of its adjacent city-region. By extension, there is no longer such a thing as 'suburbanization', understood as peripheral accretion to a centre-led urban process; edge cities may look like suburbs, but they most certainly are not. Perhaps the most pressing categorical revision in a revised city dictionary relates to the term 'sprawl'. For some, this much-maligned appellation invokes all that is bad about uncontrolled urban growth, yet for others it is the benign realization of millions of American suburban dreams, and what could possibly be wrong with that? In a new lexicon, the definition of *sprawl* as uncontrolled suburbanization is a secondary, even antiquarian usage; instead, its primary meaning would describe the principal geographical expression of contemporary urbanism. Sprawl is thus an urban theoretical 'primitive', a fundamental category for describing the way urbanism is currently being produced. Reframing sprawl in this way positions *environmental sustainability* at the core of the urban question (cf. Wolch, Pastor and Dreier 2004). Until very recently, conventional urban theory tended to relegate environmental issues to the margins, but now cities have broken their bounds, habitats have been destroyed, climates altered, species eliminated, water privatized, and biodiversity patented. The viability of life on earth is truly under threat, though this realization is desperately slow in dawning.

Our lexical revisions will also encompass urban politics and planning. Stated bluntly, urban geography has trumped local politics in the USA. By this I mean that altered geographies of contemporary urbanism are redefining the meaning and practice of urban politics. This is because the sprawl of cities beyond existing political jurisdictions negates the notion of representative democracy, compromises the ability of the local state to serve the collective interests of its constituents, and intensifies the subordination of the local state to plutocratic privatism. Today, jurisdictional fragmentation in megacity-regions has become a pathological, iatrogenic condition: i.e. the clash between urban hypertrophy and obsolete government apparatus itself causes problems and prevents the local state from meeting its obligations. This gap between institutional form and urban reality is indicative of what I call 'high modernism', a terminal condition for the institutions involved. Even politicians with a sense of accountability and responsibility are stymied by federal and state governments committed primarily to 'starving the beast', that is, denying funds for authorized public programmes and creating what Naomi Klein (2007) refers to as 'hollow government'. This task is made easier by appointing incompetent, inexperienced cronies to important office in order to ensure that the office will become dysfunctional. *In extremis*, the state and its apparatus becomes a mere instrument through which private interests plunder public wealth, what Galbraith (2008) calls a 'predator state'.

A central normative question in contemporary urban political geography pertains to the allocation of appropriate functionalities and to appropriate scales of a re-territorialized local government (Mouffe 2005, Brenner 2004). What is the optimal scale of regionalization to ensure effective representative democracy and efficient public service provision in the hyper-extended metropolis? This is, I realize, a longstanding problem that academics and politicians have never adequately resolved. But until and unless it is resolved, millions of people will remain effectively disenfranchised, un- or under-served, and even actively harmed by these 'failed local states'. In some fundamental way, the future of electoral politics, representative government, and even the possibility of local democracy seem to be threatened by the disconnect between sprawl and institutions of the local state. The rise of privatopias (including gated communities) is one example of a pathological urban outcome – an aesthetic that concretizes a perverse doctrine of anti-democratic residential apartheid, and that now controls 18 per cent of America's housing stock housing one-sixth of the nation's population. Underwritten by the politics of privatization, common interest developments (CIDs) are described by McKenzie (2005) as an orchestrated attempt to replace public municipal government with unaccountable private agencies. Only when the private overseer fails is the public sector called in to clean up the mess. Such political geographies – of privatization, secession and balkanization – are becoming a global phenomenon (Glasze, Webster and Frantz 2006). They represent profound threats to the urban polity because private cities are by definition self-interested and unaccountable; they facilitate the demise of the public realm and promote an unequal society.

Yet, paradoxically, sprawling cities also provide enclaves where intense local autonomies are possible, enabling communities-of-interest to realize their goals often below the radar of formal politics. Such movements include the activities of the much-vaunted 'creative classes' and advocates of 'green urbanism'. The potential of revitalized local social movements, globalization from below, and recovered human agency all point optimistically to a grassroots political renaissance (see Flusty 2004, Smith 2001, Keil 1998, respectively). In addition, local governments themselves sometimes find incentives and opportunities for local experiment. For instance, in Southern California, Riverside County has been attempting to manage rapid urban growth by invoking federal endangered species legislation; in Ventura County similar land-use management objectives were sought through a broadly-based coalition of grass-roots movements. Same goal, different political means. Elsewhere, cites such as Maywood, CA, declare themselves 'sanctuary cities' pledged to assist undocumented migrants. Collective local state action is also possible, as when over 700 US cities signed on to a global initiative to curb greenhouse gases. These emergent forms of socio-spatial autonomy refocus the cognitive dimensions of city dwellers and, consequently, the agenda of the urban question.

Hinterland Aesthetics

If, as I believe, urban process and urban politics are changing and urban theory requires revision, then surely our public policy prescriptions and practices also require adjustment. What *is* good public policy for private cities, spectacular urbanisms, and sprawl? On the face of it, we have reached the point where rational planning interventions suitable for the modernist city are now obsolete in the face of postmodernism's decentred, fragmented urban pastiche. If, for example, contemporary urban theory suggests that conventional downtowns are obsolete, does it really make sense to promote downtown 'renewal' when the principal urban dynamic has everywhere shifted to the periphery? Not only that, traditional corporate and philanthropic leadership has also quit the city centre, both geographically and morally, as headquarters evacuate to greenfield sites and top management is endlessly relocated, often to offshore locations. Of course, it is still possible to defend downtown revitalization on the basis of efficient reuse of the physical and social infrastructural investments already in place. However, such policy must be recognized for what it is: a hugely risky investment strategy predicted by current theory to fail. For some time, greenfield urban developments in Southern California have occurred without conventional downtowns, which are sometimes added later for aesthetic and identification/branding purposes or simply to extend consumption opportunities. Not long ago, for example, the City of Santa Ana in Orange County spent one million dollars to refurbish and repaint a water tower engraved with the plaintive claim: 'Downtown Orange County'; from the air, the Palm Springs metro region seems to be organized around a series of golf courses attached to homes for the leisured classes; and in the absence of any public square, the City of Anaheim holds many public gatherings in the Disneyland parking lot. 'Downtowns' have become externalities in this urban development process; they are no longer constitutive of the city, but merely spectacles or side-shows. This single fact is perhaps most emblematic of the altered urban dynamic behind spectacular urbanisms: cities are now being built according to new social and political-economic imperatives, including those based in diversion, perception and cognition; and the traditional artefacts of earlier cities, if they appear in the landscape at all, are simply nostalgic gestures recalling former glories.

Changes in central-city and hinterland aesthetics are indicative of a new relation between signs and practices. The cityLAB manifesto of UCLA architect Dana Cuff and her colleagues captures just how far this urban revolution has evolved:

> Transformations of the city exceed our ability to control them ... we arrive at results seemingly by fast-forward, without clear grasp of how we got there. Though not necessarily temporally fast, change occurs as a set of discontinuous jump cuts: urban development is not progressive, but it can never turn back; design is increasingly regulated without ever showing improvement.

This describes every urban professional's nightmare: a sense of a lack of control, nonlinear process, pathological outcomes, dysfunctional regulation, and aesthetic implosion. Yet Cuff's vision is fully consistent with the outline of postmodern urbanism that I have portrayed in this essay. How will urbanists imagine and act in a world of diverse, fast-forward urban aesthetics?

I readily concede that traces of a fragmented urbanism have long been present in Los Angeles and other US cities, though they may not have been universally recognized or labelled as such. Historian Eugene Moehring described the phenomenon of 'leapfrog growth' in mid-twentieth century Las Vegas, referring to the tendency for development to occur in disjointed, non-contiguous parcels. This produced what Charles Paige called a 'checkerboard effect', i.e. 'large residential subdivisions connected by commercial strips along major streets and separated by equally large squares of undeveloped land' (quoted in Moehring 1995: 238). Today, transdisciplinary comparative urban research has begun to demonstrate the ubiquity of decentred urban forms. For instance, sociologist David Halle (2003) concludes that Los Angeles highlights the changing relationship between core and periphery, a shift that is pertinent to other American cities. Geographer Richard Greene (2006) finds evidence that Chicago's modernist urbanism is presently being overlain by a postmodern scrim (also see Phelps and Wu 2009, for evidence from Beijing and Shanghai). Urban planner Klaus Kunzmann (1998) describes the emergence of 'patchwork' urbanism in the city-regions of Europe. And in their work on cities in a network society, architects Stephen Graham and Simon Marvin (2001) explicitly invoke the keno capitalism icon (fig. 2.1 above) in their world of 'splintering urbanism'.

And yet, even as the fragmented city is becoming universal, it is hard to envisage a coherent 'hinterland aesthetic' that will replace the modernist ideal of a downtown core with its collection of heroic testaments to democracy, culture and corporate wealth. For now, the hinterland seems capable only of producing a radical aesthetic flattening, a pastiche of homogeneous brands, and functions in tenuous adjacency. As I mentioned in the Palm Springs example, the hinterland-driven process is rolling out over the desert a seemingly infinite carpet of discrete golf-based leisure communities. Once in a while, spectacular urbanisms punctuate the Coachella Valley where Palm Springs is one of many urban centres – a monolithic casino hotel, a high-end retail outlet mall attached to a freeway greenfield site, and a diner to breakfast among gigantic artificial dinosaurs. People are drawn to these foci irrespective of their aesthetics, and the portfolio of hinterland aesthetics contains much that is incontrovertibly ugly. Alan Berger (2006) has described a 'drosscape', consisting of places that are part of a disjointed process of land use abandonment and (sometimes) re-use, occurring at all scales from city-regions to individual lots. In so doing, he casts original light on the dystopian urbanisms of fascination. Even the much-celebrated Las Vegas is studded with what architects Nicole Huber and Ralph Stern (2008) identify as 'landscapes of failure'.

What would a hinterland aesthetic of the obsolescent 'core city' look like? Right now, an Architecture of Serious Spectacle seems to provide an irresistible narcotic for client and designer alike, leading (for instance) to the horrifying prospect of 70-storey towers looming freestanding over Chapultepec Park in Mexico City. The Dubai urban circus demonstrates that modernist megaprojects are still a fashionable fetish among plutocrats. And Shanghai's spectacular urbanism hides the fact that one-third of its new apartment complexes are subdivided into 'collective rentals' where a single unit can house ten or more people; here architecture masks an affordable housing problem of colossal proportion. Meanwhile, in the US, long-established core cities continue to dwindle, and spectacular cultural centres seed the landscape with magical powers of urban revitalization. Elsewhere, landscapes of security and surveillance become pervasive, and privatism winks at the demise of the public realm.

If fascination represents a piquant mixture of emotional shorthand in urban cognition, it may yet become a primary vehicle for urban professionals to design new techniques of control for a mode of production based in psychosocial privatism. Hinterland aesthetics are an expression of capitalist power and socialization in our urban becoming.

The Urban Question in Postmodernity

Spectacular urbanisms are intensely personal experiences. As a consequence, they inevitably enter the urban lexicon with a surplus of meaning. They are personal and collective, material and mental, global and local, pleasurable and horrific; they encapsulate eros and thanatos; they homogenize the urban experience of rich and poor despite their unequal access. In this sense, spectacular urbanisms are, in an Althusserian sense, 'overdetermined', i.e. similar outcomes can derive from multiple causalities, and different outcomes may result from a single causality (Althusser 2001 [1971]). This complicates our analytical challenges in ways that may be irresolvable. However, indeterminacy does not grant absolution. In a world of cities and changing urban process, the need to develop better theoretical and analytical apparatuses is urgent and consequential.

The urban question in the mid-twentieth century began as a search for an urban science that could explain the rationalities of the modernist city. It evolved into a Marxist-inspired interrogation of the specificity of the urban, that is, the authenticity of a conceptual primitive (the city) that was best regarded as epiphenomenal to the capitalist mode of production. The reductionism in this categorical shift caused a third mutation in the urban question: the reconstitution of the 'city' as an analytical primitive, both constitutive of, and an outcome from, the general urban dynamic. In the cultural/intellectual maelstrom of postmodernity, the meaning and diversity of the urban experience have achieved a new prominence, powered the ubiquity of a world-wide urbanization and the rise of the network society. In my view, there is a clear correlation between the social imperatives of network society and the spatial imperatives of postmodern urbanism. I follow Debord in my desire to do harm to the society of the spectacle and its close twin, fascination, because both are implicated in the production of material inequality and difference, and of cognitive cooptation and subordination. To achieve this revision, we need to properly parse the empirical texts of the city and to develop a more robust theory capable of addressing the urban question after modernity.

References

Althusser, L. 2001 [1971]. *Lenin and Philosophy, and Other Essays*. New York: Monthly Review Press.
Berger, A. 2006. *Drosscape: Wasting Land in Urban America*. New York: Princeton Architectural Press.
Brenner, N. 2004. *New State Spaces*. Oxford: Oxford University Press.

Burgess, E.W. 1925. 'The Growth of the City: an introduction to a research project', in *The City*, edited by R. Park, E.W. Burgess and R.D. McKenzie. Chicago: University of Chicago Press, 47–62.

Castells, M. 1977 [1972]. *The Urban Question: a Marxist Approach*. London: Edward Arnold.

————— 2001. *The Internet Galaxy: Reflections on the Internet, Business, and Society*. Oxford: Oxford University Press.

Christaller, W. 1967 [1933]. *Central Places in Southern Germany*. Englewood Cliffs, New Jersey: Prentice Hall.

Davidson, C.M. 2008. *Dubai: The Vulnerability of Success*. New York: Columbia University Press.

Davis, M. and Monk, D.B. (eds) 2007. *Evil Paradises: Dreamworlds of Neoliberalism*. New York: The New Press.

Dear, M. 2000. *The Postmodern Urban Condition*. Malden: Blackwell Publishers.

————— (ed.) 2002. *From Chicago to LA: Making Sense of Urban Theory*. Thousand Oaks, CA: Sage Publications.

Dear, M. and Flusty, S. 1998. 'Postmodern Urbanism'. *Annals of the Association of American Geographers*, 88(1), 50–72.

Dear, M. and Scott, A.J. 1981. 'The Urban Question: towards a framework for analysis', in *Urbanization and Urban Planning in Capitalist Society*, edited by M. Dear and A.J. Scott. London: Methuen, 3–16.

Debord, G. 1992 [1967]. *The Society of the Spectacle*. New York: Zone Books.

Flusty, S. 2004. *De-Coca-Colonization: Making the Globe from the Inside Out*. New York: Routledge.

Galbraith, J.K. 2008. *The Predator State: How Conservatives Abandoned the Free Market and Why Liberals Should Too*. New York: Free Press.

Glasze, G., Webster, C. and Frantz, K. (eds) 2006. *Private Cities: Global and Local Perspectives*. New York: Routledge.

Graham, S. and Marvin, S. 2001. *Splintering Urbanism: Networked Infrastructures, Technological Mobilities, and the Urban Condition*. New York: Routledge.

Gray, J. 2006. 'The Case for Decency'. *New York Review of Books*, 53(12), 20–22.

Greene, R.P. 2006. 'Strong Downtowns and High Amenity Zones as Defining Features of the 21st Century Metropolis: the case of Chicago', in *Chicago's Geographies: Metropolis for the 21st Century*, edited by R.P. Greene, M.J. Bouman and D. Grammenos. Washington D.C.: Association of American Geographers, 50–74.

Halle, D. (ed.) 2003. *New York and Los Angeles: Politics, Society, and Culture: A Comparative View*. Chicago: University of Chicago Press.

Hammett, J. and Hammett, K. (eds) 2007. *The Suburbanization of New York*. New York: Routledge.

Harvey, D. 1973. *Social Justice and the City*. London: Blackwell Publishers.

Huber, N. and Stern, R. 2008. *Urbanizing the Mojave Desert: Las Vegas*. Berlin: Jovis Verlag.

Jameson, F. 1991. *Postmodernism, or the Cultural Logic of Late Capitalism.* Durham: Duke University Press.

Keil, R. 1998. *Los Angeles: Globalization, Urbanization and Social Struggles.* New York: Wiley.

Klein, N. 2007. *The Shock Doctrine: The Rise of Disaster Capitalism.* New York: Metropolitan Books.

Kunzmann, K. 1998. 'World Cities in Europe', in *Globalization and the World of Large Cities*, edited by F.-C. Lo and Y.-M. Yeung. Tokyo, New York, Paris: United Nations University Press, 37–75.

McKenzie, E. 2005. 'Constructing the Pomerium in Las Vegas'. *Housing Studies*, 20(2), 187–203.

Lefebvre, H. 1991. *The Production of Space.* Malden: Blackwell Publishers.

Lösch, A. 1954 [1940]. *The Economics of Location.* New Haven: Yale University Press.

Moehring, E.P. 1995. *Resort City in the Sunbelt: Las Vegas, 1930–1970.* Reno: University of Nevada Press.

Mouffe, C. 2005. *On the Political.* London: Verso.

Phelps, N. and Wu, F. 2009. 'From Suburbia to Post-suburbia in China? Aspects of the transformation of the Beijing and Shanghai global city regions'. *Built Environment*, 34(4), 464–81.

Pile, S. 2005. *Real Cities.* New York: Routledge.

Schmid, H. 2006. 'Economy of Fascination: Dubai and Las Vegas as examples of themed urban landscapes'. *Erdkunde*, 60(4), 346–61.

Scott, A.J. and Soja, E.W. (eds) 1996. *The City: Los Angeles and Urban Theory at the End of the Twentieth Century.* Berkeley: University of California Press.

Smith, M.P. 2001. *Transnational Urbanism: Locating Globalization.* Malden: Blackwell Publishers.

Sorkin, M. (ed.) 1992. *Variations on a Theme Park.* New York: Noonday Press.

———— 2008. 'Back to Zero: Mourning in America', in *Indefensible Space. The Architecture of the Nation Insecurity State*, edited by M. Sorkin. New York, London: Routledge, 213–32.

von Thünen, J.H. 1966 [1826]. *The Isolated State.* Oxford, New York: Pergamon Press.

Walker, R. 1981. 'A Theory of Suburbanization', in *Urbanization and Urban Planning in Capitalist Society*, edited by M. Dear and A.J. Scott. London: Methuen, 383–429.

Wolch, J., Pastor, M. and Dreier, P. (eds) 2004. *Up Against the Sprawl: Public Policy and the Making of Southern California.* Minneapolis: University of Minnesota Press.

Acknowledgements

Thanks to Heiko Schmid, Woody Sahr and other conference participants for their insightful contributions and commentaries.

Chapter 3

The City is Back (in Our Minds)

Jacques Lévy

Introduction

In this text, it is argued that, in the matter of cities as elsewhere, the 'economy of signs' is based on messages and meanings that require referents, which can be messages, too, but not the same. When we suppose there is a 'surplus of meaning', we are not speaking of pure signifiers; we are speaking of a hyper-realistic relation to the city and assuming that this relation itself is part of urbanity today. Now if we address fascination as a particular component of the subjective relationship to an environment, fascination, or magic, or enchantment, or any affective or aesthetic dimension of the city should not necessarily be opposed to awareness and rationality, but could be seen as complementary to them. Should not we define a city as a place where the surprise of discoveries and encounters is the major potential, the main productive force? Micro- or macro-events occurring in a city are not a side-effect of urban life but the very substance of urbanity. That is why what can fascinate in the city is not only specific items of a city but the city itself, the city as the unplanned programme of unexpected experience. And that is why we should place on the same level urban and anti-urban fascinations, ordinary urbanites and authorized speakers. What is at stake now for social scientists is to overcome their traditional reluctance to the concept of city, and commit themselves to assess urbanity without prejudice as a heritage and a process.

If we try to make ourselves available to this serendipity, i.e. the fact that unexpected discoveries bring a fundamental contribution to the comparative advantage of a city, we should first carry out a stern, swift criticism of anti-urban stances in different kinds of texts, including social science corpuses.

Then it will be easier to address peculiarities of the circulation of signs in urban environments: the importance of bodies, viewed from the point of view of both motivity and sensoriality, as operators in public space and as key indicators of the level of otherness urbanites wish for, fear, or accept. The special efficiency of urbanity for social cohesion as well as for economic productivity suggests a final aspect of urban fascination: its radical and uncompleted modernity.

What is at stake now for social scientists is that they should overcome their traditional reluctance to the concept of the city. The goal of this text is thus to contribute without prejudice to an assessment of urbanity as a heritage and a process. It is argued, too, that, in the matter of cities as elsewhere, the 'economy of signs' (Baudrillard and Levin 1981) is based on messages and meanings that require referents. These referents can also be messages, but not in the same semantic frame. Signs always mean something and the hypothesis of a pure signifier was not confirmed by experiment. When we suppose there is a 'surplus of meaning', we are not speaking of pure signifiers; we are speaking of a hyper-realistic relation to the city that is a possible illusion of transparency between the medium and the message. We can accept the idea that this is part of urbanity today, but more generally of social reality. But in contrast the surplus of reflexivity provided by the emergence of actors in a society of individuals (Beck, Giddens and Lash 1994, Elias 1991) largely compensates for the fallacy of an immediately readable world.

Negative Archaeology

Why is it so difficult to simply describe what happens in a city? Why is everything that happens in a city because it is a city so often devalued? Placing on the same level urban and anti-urban fascinations, ordinary urbanites and authorized speakers allows us to present the pervasive presence of anti-urban stances in different kind of texts, including social sciences corpuses.

The Anti-Urban Fascination 1: Ideologues

The history of the city entered our minds through a spell. The curse on the city (and indeed on globalization), announced in the Old Testament was maintained and given fresh impetus by an important European Puritan-communal tradition.

> 1 Now the whole Earth used the same language and the same words. 2 And it came about as they journeyed east, that they found a plain in the land of Shinar and settled there. 3 And they said to one another, "Come, let us make bricks and burn them thoroughly". And they used brick for stone, and they used tar for mortar. 4 And they said, "Come, let us build for ourselves a city, and a tower whose top will reach into heaven, and let us make for ourselves a name; lest we be scattered abroad over the face of the whole earth". 5 And the Lord came down to see the city and the tower which the sons of men had built. 6 And the Lord said, "Behold, they are one people, and they all have the same language. And this is what they began to do, and now nothing which they purpose to do will be impossible for them". 7 "Come, let us go down and there confuse their language, that they may not understand one another's speech". 8 So the Lord scattered them abroad from there over the face of the whole earth; and they stopped building the city. 9 Therefore its name was called Babel, because there the Lord confused the language of the whole earth; and from there the Lord scattered them abroad over the face of the whole earth (Genesis 11: 1–9).

The myth of Babel clearly poses an antinomy between a God-provided milieu and a self-built human environment. The city is used as a metaphor to privilege transcendence against historicity, *Gemeinschaft* against *Gesellschaft*, command against design, morals against ethics.

This tradition is represented at the highest philosophical and literary level by Jean-Jacques Rousseau (1960 [1761]: 251), who wrote: 'The major inconvenience of large cities is that people there become something other than their real selves, and that society gives them, as it were, beings different to their own.'[1] Such a tradition was especially strong in Europe during the formation process of societies in the Nation state, whose ideologies usually referred to more rural and communitarian principles. It was enhanced by 'agrarian' states, which perceived emerging large cities as dangerous political rivals. This was particularly the case in France, where the alliance between the central state and rural 'notables' kept Paris and other major urban areas in a state of subjugation (Marchand 1993). In many European countries, the 'imagined community' (Anderson 1983) that underpinned the national identity referred to the myth of a self-sufficient, nature-friendly, prolific countryside harmony which was unfortunately undermined and threatened by a messy, artificial, and sterile city (Salomon-Cavin 2005). This myth migrated to America, conveyed by people who, from the day they landed, began to build an overwhelmingly urban environment, while remaining proud and ardent militants of the Puritan cause. This probably explains a paradox: the North-American version of the anti-urban narrative is embodied, if not in a city, at least in a landscape of towns, that of 'Main Street USA' with its single-street 'parochial'[2] community and its farming hinterland, but not exactly in the plough, ploughing and the ploughman, as was the case in European mythologies. In Walt Disney's images or in the Henry Ford Museum in Dearborn, Michigan, the discourse may seem to have been secularized. However, when one reads the following statement in a Dearborn-published article by Henry Ford, stigmatizing the 'pestiferous growth' of the city, one discovers an unquestionable continuity with the Bible's inspiration and style.

> The city is doomed. ... The modern city is the most unlovely and artificial sight this planet affords. The ultimate solution is to abandon it. We shall solve the city problem by leaving the city. ... The suburban car and the automobile have rendered confinement within the City unnecessary for large numbers of people. Get people into the country, get them into communities where a man knows his neighbor, where there is a commonality of interest, where life is not artificial, and you have solved the City problem (Ford 1922).

---

1   'C'est le premier inconvénient des grandes villes que les hommes y deviennent autres que ce qu'ils sont, et que la société leur donne, pour ainsi dire, un être différent du leur.'

2   This is the word Lynn Lofland (1998) coined to designate the kind of fake public space where anonymity is impossible.

The Anti-Urban Fascination 2: Scholars

With Heidegger's stance, the verdict is cast again, this time not in an ordinary language but in a scholarly register. There is no theory of the city in Heidegger's works, only a general approach of space (Heidegger 1971) as a component of the human existence, called *Wohnen* [dwelling]. But the awareness of the *Wohnen* supposes *Bodenständigkeit* [rootedness] and rejects *Neugier* [curiosity], *Gerede* [gossip], or *Zweideutigkeit* [ambiguity], all *urban* attitudes that lead to an 'oblivion of the being'. While it may be somewhat derogatory for the philosopher, I would argue that Marshall Philippe Pétain, the fascist guide of the *État français*, was the best pupil of Heidegger's teaching (though he must have been unaware of it), when he repeated his creed about territorial planning and social policy: *La terre ne ment pas* [Land does not lie].

Among French-speaking, neo-Heideggerian scholars in 'urban philosophy' such as François Choay (1993), an author who remains an important thinker for the philosophy of space,[3] a common statement is that the city has vanished, been buried, surrounded, devastated, privatized, made obsolete or 'museumized' by the urban sprawl. In a sense, these authors, who lament the end of 'traditional' urban culture, make the same diagnosis as their supposed opposites, authors who joyfully celebrate the end of the city, such as the libertarian utopia epitomized by Melwyn Webber (1964) or the *cybermythology* proposed by 'early' William Mitchell (1995). They are not so far either from those such as Thomas Sieverts (2003), who agree on the prophecy and package it up in irony, good will or fatalism, respectively. 'Fascination' would be the way to give an aesthetic colour the same stance. Koolhaas' Lagos (2001), featured as the epitome of the global urban future, is the smiling counterpart to the sad generalization Sieverts incautiously creates by confounding the entire urban world with his playfields in the *Ruhrgebiet* Rust Belt.

Geography in Outer Space

Geography has not been a prominent player in the revival of the concept of city. And this is all the more true in North America. In this regard, Ed Soja's 'putting the city first' (2003) can be seen as a rare exception. His peculiarity in the literature comes not so much from his subtle analysis of various 'species' of urbanity in a context (the Los Angeles area or its clones), where urbanity is unlikely to be met, but rather from his decision to place his whole framework under the patronage of both Jane Jacobs and Henri Lefebvre. This companionship reveals Soja's reading of Lefebvre: not as a freshly re-wrapped, old-fashioned Marxist, as most North-

---

3 See Heidegger's seminal lectures 'Building, Dwelling, Thinking' and 'Poetically Man Dwells' (both in Heidegger 1971). For a talented contemporary re-reading of Heidegger's conceptions of space see Peter Sloterdijk (2004).

American social scientists see him, but as a useful transitional object towards an emerging approach capable, at last, of taking the city as such seriously.

In general, geographers have let urban thought slip out of their hands simply because they have failed to declare themselves interested. In Germany, urban space was leased to sociologists, in Britain and France, to nobody. As an heir of its German counterpart, the Chicago School, in the early twentieth century, embodied various spatial dimensions of the urban sciences: quantitative studies carried out on the basis of a behaviourist and positivist approach; qualitative surveys inspired by exotic anthropological inquiries; theoretical works taking advantage of the original link between sociology and philosophy. In this category, Louis Wirth epitomizes the fertility of the encounter between Simmel's intuitions or reflections and empirical field studies in an lively North-American metropolis.

Why, for instance, do North American geographers seem so uninterested in the new trend emerging in North American cities? Will the scholars be the last to notice that the city *is* intrinsically a revolution? And that taking up urbanity is not a peripheral item but a major event? The following answers to these questions are deliberately simplistic, in an attempt to identify unsolved problems:

1. Geographers are not interested in cities. They prefer rural areas, even if the entire world is urban.
2. Geographers have yet to bridge the gap with post-structuralist social theory. They continue to see society as if it were devoid of human beings. They see social dynamics as driven by faceless structures such as 'capital', 'African-Americans', or 'women'. Is geography condemned to remain trapped in a pre-pragmatic turn? The theoretical and empirical exploration of contemporary historicity is dramatically absent, replaced by a standardized, politically-correct newspeak.
3. Geographers were not, in the past, interested in politics, only in techniques. Now they remain reluctant to recognize politics for what it is, namely: the controversial construction of a methodology capable of convincing the members of a society that violence is not the solution to achieve their private goals or (the same thing put in another way) that living in a society managed by such a method generates for its members a positive-sum game. On the contrary, most geographers imagine politics as a mere confrontation of particular interests, namely a shock between various communalisms, as if it were geopolitics: a zero-sum game. In practice, they prefer to look at street wars than to consider the role of urban space as a political issue in itself, even less as a potential peacemaker.
4. English-speaking geographers are no longer interested in space. They prefer speech upon speech upon speech on space. Opposing constructivism to realism instead of articulating them is a severe hindrance for the social sciences.
5. American scientists, like other Americans, hate the city … even if some of their compatriots are currently changing their mind on this matter.

6. Many intellectuals do not like society when it succeeds. Eschatological pessimism is for them a professional ideology and a power tool. To the contrary, the city is, in itself, the evidence for a possible societal being-together.

It may be argued, of course, that these weaknesses are not exclusive to geographers. A non-geographer scholar like Mike Davis, for instance, could be criticized because of his disaster-oriented view of cities, and his motivations are probably linked to points V. and VI. Epistemologically and theoretically, however, geography is the missing link that has proved difficult to replace. This is probably due to a simple fact. A city is a spatial reality from whatever scale we try to address it, from the local to the global. In principle, the city is a spatial choice societies have made in order to manage the question of distance in a particular way. Co-presence (the maximum of realities in a minimum of extension) is in competition (I should say, more precisely, in *co-opetition*) with two other kinds of solution: mobility and telecommunication. If we fail to appreciate its spatiality, we are bound to understand virtually nothing in a city. We cannot understand it because we observe the consequences of a very peculiar spatial arrangement and try to relate them to the non-spatial. The resistance of social, theory-oriented geography to 'spatialism' and to the myth of 'general laws of space', which were mere laws of geometry, served a useful purpose. But, conversely, in the same context, nothing can justify the refusal to accept any explaining power of space, if space is defined as a completely social component of societies. Both 'spatial analysis' activists and anti-spatial militants of single-principle economic, sociological, or political paradigms had good reasons to reject spatiality as a fully-fledged dimension of the social world. The concept of city was one of the numerous collateral victims of this deliberate ignorance.

Will geographers finally meet the city? In this regard, the gentrification issue is more than simply a good example; it is somehow a crucial experiment. If it is said that the rich should live together with the poor, which could be a useful leverage for social mix, some geographers reply: 'Nobody would accept this.' If it is observed that the so-called *bobos* ('bourgeois-bohemian') pretend that they are in favour of it, the same people reply: 'They say it but they won't carry it out. And if it is argued that this actually occurs, they reply: "They do it for the wrong reasons."'

Social Justice and the City (1973) is probably a pivotal book among David Harvey's works. It is his last attempt to develop, on the basis of Marx and Engels' thought, a rational construction compatible with his previous epistemological inquiry, *Explanation in Geography*, published in 1969. We can perceive in *Social Justice and the City* a 'theoretical pain', whose message would be: I have done my best to create a realistic Marxist theory of the city; now, if I want to continue to be a Marxist, I will have to adjust the urban reality to my postulates. Harvey identifies actual theoretical difficulties, e.g. about the meaning of the land rent issue in an urban context, but finally decides to get rid of them by carpet-bombing the question marks with safe ideology on the working class, exploitation and revolution. As one of Harvey's spiritual sons, Neil Smith (1987) has agreed to carry the burden. His debate with Chris Hamnett (1991), however, is reminiscent of a dual-level road system in which there are very few interchanges. Hamnett's arguments are simple and difficult to challenge, except for when an implicit axiom comes to the surface: the first non-poor, non-ethnic person moving to a formerly 100 per cent poor, ethnic area is by definition committing an act of violence against the working class. Here lies the paradox: the denunciation of gentrification is not made for the sake of social mix but, to the contrary, in the name of an enhanced ghetto. Either the concept of urbanity as consisting of dense diversity is implicit but obvious (Hamnett), or it is firmly rejected as a dangerous illusion (Smith). In these circumstances, the characters of this play are rarely together on the same stage. In this sense, many North-American geographers, who had fiercely denounced the ghettos as a product of capitalism, are denouncing even more strongly the possibility of overcoming them by efficient policies of social mixing.

Would Harlem's revival as a multifunctional, vibrant neighbourhood be worse than its decadence as a ghetto? More generally, they almost completely neglect the major historical event of 'Urban Renaissance', which is happening in North America, changing the social landscape of virtually all inner cities. They are missing this event, because they are still wearing their dark glasses that prevent them from accepting phenomena which are, unfortunately, rebel to their cause.

Figure 3.1   Harlem 2003 – still a poor and unattractive neighbourhood

Thus, a strange 'coalition' is appearing between those who hate cities because they hate the idea of an interdependent, solidarity-based society, which is actually the quintessence of *societalness*, and those practising a radical *Selbsthass* [self-hate] hate the city because they hate the idea that their own ideal, instead of the

Figure 3.2   Harlem 2003 – social and ethnic mix still to come, but some features of a renaissance are already appearing

working class, might be a good blueprint for a desirable city. Fascination is the translation of this alliance into the contention of a de-realization of urbanity: if what is called urbanity is nothing but a superficial marketing envelope for an impossible dream, then let's denounce simultaneously this superficiality and those who pretend to offer more.

Figure 3.3   **Chicago 2003 – the comeback of the nineteenth–early twentieth century inner city means an increasing social mix, a revival of public transport and a new focus on public space**

Beyond the City of Angels: City of Bodies

In contrast, what is, for me, fascinating in the city is the fact that the concept and the reality have both efficiently resisted half a century of urban turbulences. Forty years after they were first published, Lefebvre's works continue to attract considerable interest because they are not easily reducible to a banal discourse. Lefebvre admitted that the urban link, in other words the kind of social interaction cities generate, is from an economic or political point of view, a weak link. Nevertheless, this weak link is presented as a key force in the shaping of societies. Architects and political utopians have found it difficult to reconcile this apparent antinomy.

A city is not a masterpiece and urban planners are neither authors nor artists. First of all, they have to acknowledge that they are just one set of players among others. Secondly, if they want to be efficient, while they might not necessarily have to obey other stakeholders, they need to at least listen to them. A city is not imaginable on the model of the formal blueprint of an ideal society. It is neither a law nor a constitution. Its political construction encompasses the overall complexity of civil society. If this complexity is shattered, urbanity is over. This weakness appears in all classical urban utopias, from Plato to Thomas Moore or Charles Fourier: in these discourses, the city is supposed to be the pure embodiment of the ideal construction, from scratch, of a new society. We can see with new towns how difficult it is to 'create' a city. A city is always an ex-post reality.

Cities cannot be summoned to solve all the problems of a given society. There is a slight misunderstanding here. Asking cities to do what they cannot and then concluding that we have bad cities is unfair. By themselves, cities cannot eliminate poverty and illiteracy, or protect us from war and exploitation. If we believe, however, in the self-perfectibility of societies, we can and should ask cities to make their intrinsic contribution to a desirable evolution. Roughly speaking, this contribution can be seen in terms of an increased productivity of human interaction. This comes from a concentration of diversity, allowing for a more efficient management of inputs and transforms the broad exposure to otherness into a creative device. Its usefulness is verifiable in terms of economic growth but also in terms of social cohesion, political governance, and environmental protection. For these reasons, urbanity is undoubtedly a dimension in any consistent project of a democratic republic based on justice. Urban links are weak just as 'weak interactions' in physics are weak: less dramatically, but more decisively.

As Marxists, the early Manuel Castells, the author of *La Question Urbaine* [*The Urban Question*] in 1972, and David Harvey, stumbled across the city because they tried to encompass it within 'ultimate-principle' thought. Reading Lefebvre seriously, it can be understood that there is a strong ethics and aesthetics in the city. And, as Lefebvre reminds us, there is even room for utopia: the creation of an ex-post utopia, in which a zero-dimension space emerges from the very existence of a banal city, capable of achieving the Greek ideal of *synoikism* (a place to be-together).

We have to go back to basics, or rather to the unique moment in the history of the social sciences when sociology detached itself from philosophy at the turn of the twentieth century. The connection between the maverick philosopher of modernity, Georg Simmel, in good company with Ferdinand Tönnies, Max Weber, and Karl Mannheim, is clearly present in its American projection, the Chicago School. Louis Wirth's *Urbanism as a Way of Life*, published in 1938 or, with a more economic inspiration, René Maunier's *The Definition of the City*, written as early as 1910, are excellent syntheses of what the best non-spatial but open, cultured social sciences could say 70 to 100 years ago about urban space. That is why these key texts can still be used as the first step in the training of urban scientists, including geographers.

Spatiality and *societalness* are the two necessary preconditions for a new beginning for a concept of the city. At the moment, spatiality is a major rationale of society-making and scientists who have integrated this approach into their work provide us with the bases for complexity-oriented urban studies.

City as a Co-Presence of Productive Bodies

The city is a way of generating a whole society thanks to the principle of co-presence. From the Neolithic age on, there have only been three major means to manage distance, one of which is co-presence. Many observers have predicted the end of co-presence because of mobility or telecommunication. But the three modes seem to be in a situation of *co-opetition*, that is to say a compound of co-operation + competition. Each of these rationales uses the others for its own objectives but is also used by the others. There can be no city without telephones and internet. No email without places. No e-commerce without a delivery system. The key aspect of co-presence is probably the involvement of the body, whereas mobility and above all tele-communication can get by without it. That is the reason why, in spite of the dramatic increase in the number of simulation and role-games on the internet, there is something which tele-communication cannot bring (and this is simultaneously its advantage and its drawback): a banal exception, a weak link, a hazard and a promise that an ordinary public space can provide.

Table 3.1   Three Ways for Managing Distance

	Tele-communication	Mobility	Co-presence
Accessibility to identified information	+	=	–
Serendipity	–	=	+
Exposure to place otherness	–	+	–
Exposure to body otherness	–	=	+
Type of space generated	*Place, Network*	*Network, Territory*	*Territory, Place*
Examples	*Mail, Books, the Web*	*Intercity and Intra-City Links*	*City, Home*

Source: Lévy 1999: 237, modified

In this perspective, public space should occupy a core function in any urban theory. Ulf Hannerz (1980), Lynn Lofland (1998), and Isaac Joseph (1998) are good examples of how we can take advantage of anthropology and sociology in this regard. A public space is a place inside an urban area, where everybody knows than he/she can expect to experience an urban diversity whose magnitude might be similar to that of the overall urban area.

Public space is a piece of actual, material as well as immaterial, space. It should not be confused with the public domain, that is: state property, or with the public sphere, a notion from political philosophy and translation of the German expression *Öffentlichkeit*.

Public space is 'a reasonable utopia' because there is nothing impossible in its principle, except for the risk that some of the population might refuse co-presence with others. A public space is a fundamental and fragile expression of urban society. It is a place where what is called *civility* is practised. Erving Goffman described the way people are supposed to look at others as a type of 'civil inattention' (1984 [1959]). Civility has unwritten rules, and therefore conflicts over such rules are particularly complex. For instance: can I stop in the middle of the pavement, can I sit on the grass, can I whistle if I see a beautiful girl or boy, can I read someone else's paper over their shoulder as we sit on the same bench...

Serendipity and Sensorial Apparatus

When you are *not* in a cinema, you are not the same person whether there are 1 or 300 films to see. This is *virtuality*, which should not be confused with simulation or approximation. A large city is characterized by the importance of its virtuality: there are so many things to do you have to choose, to eliminate. The more you live in an urbanity-rich environment, the larger the gap between what you could do and what you actually do: the golden burden of urbanity. In this context, chance offers a partial solution. The non-actualized part of the potentials offered by the city is made available by random encounters. *The Three Princes of Serendip* is a medieval Persian tale, popularized by Horace Walpole in the eighteenth century. Princes in the story use two alternative strategies to get through difficult situations. The first is to take advantage of unexpected things they meet along the way, while the second is to employ a capacity to arrange 'accidents' (an expression of their 'sagacity', to use Walpole's words). In an urban context, we can then identify the conditions to be met for efficient serendipity:

1. The activation and involvement of all our perceptive captors, indeed our complete sensorial apparatus, in order not to miss information. Pedestrian metrics are the most suitable for that purpose, although they are challenged by physical or digital libraries, and even more so by research engines.
2. An exposure to otherness that supposes that otherness exists (diversity) and is accessible (density), and of course that one accepts such exposure.
3. The cognitive skill to usefully interpret the information one receives.

Thus, reciprocal reinforcement takes place between urbanity and serendipity: serendipity is all the more necessary because urban space generates a rich set of virtual resources; at the same time, serendipity is all the more possible because urban space is dense and diverse. This is a clear case of virtuous positive feedback. Conversely, if we diminish the level of urbanity, we weaken the efficiency of serendipity as well as its usefulness.

Serendipity is fundamentally based on the presence of bodies available for a multi-sensoriality cognitive experience. The power of serendipity is thus one of the most fascinating fascination devices in a city.

The City: An Experience of the Human Non-Finitude

What finally should fascinate us it that there is no size limit to urban overproductivity. A classically lazy statement is that there is a critical size for metropolises. In almost all states, large cities have productivity levels (gross urban product/inhabitant) superior to the national average. The largest of all, Tokyo, is also one of those where each dweller is economically most efficient. Even cities that look smaller enjoy a scale economy because they are connected to a network of cities. Today the hinterland of any city is the archipelago of worldwide cities.

Lifetime and lifeline contexts are important in defining the relevant scale. On the other hand, no relevant limit of a 'human-size city' has been discovered. What can be observed, however, are unequal capacities in terms of creating a local or regional context where an individual can effectively reconcile his/her identity with the necessary resources for his/her development. In the case of Tokyo, more than 30 million people live in the same local society, with a common technical and anthropological environment, a situation that cannot be found either in Europe or in America.

What is fascinating in a city is the simplicity of its modernity, the simple reasons that put it everyday more at stake in the contemporary control panel of strategic political choices. When Lefebvre speaks of 'urban revolution' (2003 [1970]), he clearly means that the city is in itself, a revolutionary agenda. Why? Simply because, urbanity supposes that all the urbanites are convinced they are playing a positive-sum game by the very act of experiencing the urban space. The urban part of our societies is a contribution to a post-corporatist, post-capitalist, and post-communal society, what Norbert Elias (1991) called a society of individuals. Thus, the revolution of urbanity is not for tomorrow, but for today, in public transportation schemes, public space refurbishing, and social mix policies. Urbanity does not solve social problems but it contributes to the emergence of a social world in which social problems can be solved. As a completely artificial object, the city is an image of the power and the limits of what a society is able to do with itself. It is an experience of human non-finitude.

A City is a Place Where You Can Vote with Your Feet

The city is both in bodies and in minds as well. In fact, what we call *real* includes the infinity of its discourses and images, which are produced through the same mechanism and logics that produce its materiality. A vote, for instance is at the same time, 'real' and 'representation'.

Those that believe, wrongly, that everything is political and those that are convinced, rightly, that there is politics in everything should look thoroughly at this map (fig. 3.4). It shows that there is a strong relationship between anti-otherness vote and voters' periurban or exurban location. This referendum about religious freedom generated roughly the same map as other referendums where membership of the European Union (1992 and 2005) or acceptance of homosexual partnerships (2005) were at stake. In this period, in Switzerland and elsewhere in Europe or in North America, cities, namely central cities, and especially centres of big cities showed a continuous openness to various kinds of exposure to otherness.

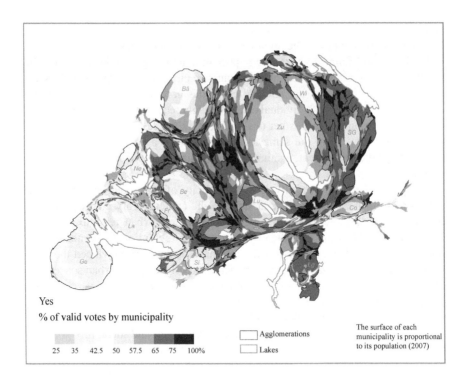

Figure 3.4   Election results of the Swiss referendum (popular initiative) on the banning of the construction of minarets, November 29 2009

Source: Chavinier and Lévy 2009

Anti-urban fascination for urban fascination is not so fascinating. At least, let us not be more fascinated, and therefore stunned, than the subject of our studies, i.e. the ordinary individuals that make the city. They are able, in many cases, to distinguish the superficial from the fundamental, the superfluous from the essential. It can be understood now that the title of this chapter should end with a question mark: 'Is the city back in our minds?' And my answer would be: let the city *as such* once and for all come into social scientists' minds. If it does not, we might miss the spatial component of social theory.

References

Anderson, B. 1983. *Imagined Communities*. London: Verso.

Baudrillard, J. and Levin, C. 1981. *For a Critique of the Political Economy of the Sign*. St. Louis: Telos Press.

Beck, U., Giddens, A. and Lash, S. 1994. *Reflexive Modernization: Politics, Tradition and Aesthetics in the Modern Social Order*. Cambridge: Polity Press.

Castells, M. 1972. *La Question Urbaine*. Paris: Maspero.

Chavinier, E. and Lévy, J. 2009. 'Malaise dans l'alteridentité'. *EspacesTemps.net*. [Online, December 2009]. Available at: http://espacestemps.net/document7961. html [accessed: 29 March 2010].

Choay, F. 1993. 'Le règne de l'urbain et la mort de la ville', in *La ville. Art et Architecture en Europe 1870–1993*, edited by J. Dethier and A. Guiheux. Paris: Centre Georges Pompidou, 26–35.

Elias, N. 1991. *The Society of Individuals*. London: Basil Blackwell.

Ford, H. 1922. *My Life and Work*. Dearborn, MI, self-published.

Goffman, E. 1984 [1959]. *The Presentation of Self in Everyday Life*. Harmondsworth: Penguin.

Hamnett, C. 1991. 'The Blind Men and the Elephant: The Explanation of Gentrification'. *Transactions of the Institute of British Geographers*, 16(2), 173–89.

Harvey, D. 1969. *Explanation in Geography*. London: Arnold.

————— 1973. *Social Justice and the City*. Baltimore: The Johns Hopkins University Press.

Hannerz, U. 1980. *Exploring the City: Inquiries toward an Urban Anthropology*. New York: Columbia University Press.

Heidegger, M. 1971. *Poetry, Language, Thought*. Translated by A. Hofstadter. New York: Harper Colophon Books.

Joseph, I. 1998. *La Ville sans Qualités*. La Tour d'Aigues: L'Aube.

Koolhaas, R., Boeri S., Kwinter S. and Tazi, N. 2001. *Mutations*. Barcelona: Actar.

Lefebvre, H. 2003 [1970]. *The Urban Revolution*. Minneapolis: The University of Minnesota Press.

Lévy, J. 1999. *Le Tournant Géographique*. Paris: Belin.

Lofland, L.H. 1998. *The Public Realm*. New York: de Gruyter.

Marchand, B. 1993. *Paris, histoire d'une ville: XIXᵉ-XXᵉ siècle*. Paris: Seuil.

Maunier, R. 1910. 'The Definition of the City'. *American Journal of Sociology*, 15(4), 536–48.

Mitchell, W.J. 1995. *City of Bits. Space, place, and the Infobahn*. Cambridge, MA: MIT Press.

Rousseau, J.-J. 1960 [1761]. *La nouvelle Héloïse*. Paris: Garnier frères.

Salomon-Cavin, J. 2005. *La Ville mal-aimée*. Lausanne: PPUR.

Sieverts, T. 2003. *Cities without Cities: An Interpretation of the Zwischenstadt*. London: Spon Press.

Soja, E. 2003. 'Writing the City Spatially'. *City*, 7(3), 269–80.

Sloterdijk, P. 2004. *Sphären, 1. Blasen, 2. Globen, 3. Schäume*. Frankfurt am Main: Suhrkamp.

Smith, N. 1987. 'Of Yuppies and Housing: Gentrification, Social Restructuring, and the Urban Dream'. *Environment and Planning D: Society and Space*, 5(2), 151–72.

Webber, M. 1964. 'The Urban Place and the Non-Place Realm', in *Explorations into Urban Structure*, edited by M. Webber et al. Philadelphia: University of Pennsylvania Press, 19–41.

Wirth, L. 1938. 'Urbanism as a Way of Life'. *American Journal of Sociology*, 44(1), 1–24.

Chapter 4

Emotions in an Urban Environment: Embellishing the Cities from the Perspective of the Humanities

Jürgen Hasse

The postmodern aestheticization of cities constitutes a *dissuasive*[1] form of socialization, so that it is not arguments but immersive suggestions that influence city dwellers' activities. Although it is customary in the social sciences to identify aesthetics (insofar as it concerns theory) with images, and thus to 'decode' analytically the iconological description of physical urban space within a semiotic perspective it would seem that such an analysis is inadequate. Built images are more than meanings; they have a scenographic character that can only be fully experienced in living movement. Therefore, 'beautified' cities are more directed towards all-encompassing emotions rather than towards simple idiosyncratic sensations. Insofar, these emotions communicate an overall impression expressing something that is more a general reflection of the *Zeitgeist* (cf. Kluck 2008) than the result of a specific idea, and thus the perception of public space derives less from singled-out architectural details than from a complex and indissoluble *Gestalt*.

The following essay addresses the question of the logics that underlie the aesthetic staging of cities (principally in metropolises), investigating the means that are used to communicate scenes of beauty and how they affect the experiencing individual. Seen against the background of the epistemologies of the social sciences, emotions may be ontologically 'non-objects' of investigation, but I shall attempt to demonstrate that the often hidden knowledge of feeling informs common 'social actors' on how to create, modify or react to a milieu, and thus makes social use of that diffuse experiential knowledge of imprecision that is the result of the suppression of rationalization processes and its subsequent discursive discipline, to which the autonomous experience of the modern subject is subjected. Experts on the creation of emotional situations such as designers, light artists, or landscape architects, therefore act on an epistemological basis that is difficult to describe, as this kind of knowledge is anchored more in experienced life-worlds. This contrast between 'scientific' ontology and lived experience makes for

---

1 The concept of 'dissuasion' as used here and in the following text is to be understood in the sense of Jean Baudrillard (1983) – as a gesture of radical seduction, with the suggestive character of a gift.

political controversy, because it reveals that emotions (at least implicitly) are also at the centre of economic as well as political geography, making them an essential medium for a different type of significant communication.

The Postmodern Glamour of Cities

The wave of aestheticization that has affected European metropolises since the 1990s (starting and beginning some ten years earlier in the USA and UK) has permanently changed the image of urban postwar modernism, mainly in the city centres. The aesthetic attempt to upgrade the atmospheric quality of urban experience is the result of a number of different, and above all economic and political, interests, involving also cultural currents, so that the renewal of aesthetic forms can not generally be ascribed to any specific interest. Instead, the significance assigned to plazas or the façades of public and private buildings is, like fashion, subject to an autopoetic regulation arising from the *Zeitgeist* of mature post war consumerism.

Until the 1970s and 1980s, classic consumer interests were directed towards assuring basic necessities. From then on, they gradually expanded to include so-called 'postmaterial values' (cf. Abramson and Inglehart 1995). The demands of a growing middle class for recreation and new experiences implied aesthetic expectations directed both to private and public spaces, resulting from an introverted, narcissistic subjectivation process in the configuration of the postmodern subject. Successively introduced sociopsychological transformation of individuals passed through situational evaluations that are gradually converging into equivalent dispositions of emotions. This kind of socially induced change in the experience of the self and the world implies a tendency towards depoliticization. An attitude of 'openness' (in the sense of indifference) has come to respond adequately to these incommensurable appearances. In the wake of this development a pragmatic, superficial culture of valueless relativism spread. Aesthetics became a medium of compensation for indifference, reinforcing the desire for (self-)experience, and thus the general aestheticization of life appears as a medium of distinction.

With the end of the 'Soviet Block' the last pillars of a utopian social model based on the ideology of 'justice' collapsed. Consequently, the new global reconfiguration of markets has brought not only private companies, but also municipalities under increased political and economic pressure to promote themselves in a competitive scenery, and the removal of trade restrictions through globalization has created both opportunities for growth but also new limitations. So metropolises also began to compete in the immersive field of aesthetics, both to realize new options and to neutralize restrictive potentials, and firms and public institutions (at first predominantly in the West) strove to demonstrate cultural superiority through the medium of images. Thus, excentric and affective gestures of construction, remodelling, transcription, and scenographic reprogramming expressed themselves in (a) presentation and (b) representation.

In this context, the logic of *presentation* is semiotically open and permits a broad interplay of cultural association, collage, and performative adaptation. Its medium is the fascination of experience, which is self-contained *as form*. Here, private companies and public institutions participate in similar ways in measures to improve the presentational character of cities (their qualities of location). Such a competition is responsible for the aforementioned process of depoliticization, which is increasingly affecting the postmodern world. Prettified cities gain their identity through their pleasant atmospheres (cf. Müller and Dröge 2005: 101), a fact which is largely unrelated to any search for meaning. Under the auspices of a post-critical constitution of the subject, the apparatus for the regulation of emotions via the culture industry is spreading. So, finally, postmodern aestheticism is filling the vacuum of moral values and political programmes, leading to an experiential loss of rational comprehension and a vanishing trust in the effectiveness of local, democratically legitimated, power structures differentiated throughout a field of widely varying subcultures. Such a situation was rendered tolerable through affect-logical (Ciompi 1988) dissuasion in the light of beauty through the presentation of aesthetically arranged worlds of compensation.

Representation follows a different logic, much more linear, than *presentation*. This logic communicates proposals of significance, which can be of an economic, political, or cultural nature according to their systemic function. So 'representative buildings' of private companies communicate an excess of meaning, playing with magic insofar as they suggest credibility, gravity, power and trustworthiness. In a structurally similar manner, major state institutions exploit the same means of 'radical seduction' in their architectural staging. While the aesthetic interests of companies are economically coded, public institutions attempt to inculcate a feeling of identification, which is subordinated to the authority of a sovereign state – not through political compulsion, but rather through the dissuasive effects of fascination. Thus in a period of economic culture or cultured economy, the city represents an aestheticized image of different lifestyles, whose depoliticizing content expresses itself in the (beautiful) aesthetics *of forms*. Therefore, for example, the giant advertisements that cover the façades of construction sites are not only pictures of fashion, they implicitly also act as an anaesthetic to the dirty and provisional environment where they have been posted. The pure glamour of the apparently aseptic city becomes a standard for a generalized aesthetic expectation of perception and experience, and, subsequently, the physical cleaning of the city thus reflects in one of its deep effects the post-critical cleaning of the regime of perception. Following Welsch (1993), we can speak of an aestheticization of surfaces and depths that form a complex dynamic interrelationship. The perception of aesthetic forms, figures, and situations of representation can not be reduced either to vision or to symbolic understanding, but is deeply rooted in the *Wahrnehmungs- und Aneineignungsweise* [ways of perception and appropriation] (Müller and Dröge 2005: 100) which build up patterns of relationships that regulate individual and social life with respect to the economy, politics, and culture.

Thus, the aesthetic-suggestive 'foundation' of reality expresses the logic of an aesthetic 'dispositive' in Foucault's sense (cf. 1978 [1977]). In this context, the current offensives for aesthetization communicate secret arrangements of non-verbal (but nonetheless discursive) powers, contributing as the 'dominant tendency of contemporary cultural development' (Müller and Dröge 2005: 95)[2] to a universal aestheticization of social life, a social life that progresses most effectively when remaining largely unconscious beyond the spoken word and rational alternatives for action. Here, it is not reason but emotion which is addressed. In this sense, the experience of material things (city halls, company headquarters, court buildings, bridges, etc.) becomes a sublime fascination, as happened with the World Fairs at the turn of the nineteenth to the twentieth Century, which served culturally to communicate fascination through technical superiority. Their 'superiority' dissolved the barrier between rationality and irrationality through the ambivalent feelings where the unity of the beautiful *and* the hideous, of fascination *and* abhorrence appear together (compare Kant and Burke's treatment of the superior with Lyotard 1989).

Understanding Aesthetical Spatial Experience – Views from Outside Geography

The *experiencing* of a spatial environment is characterized by *emotional* participation. In its pathic form, such living experience is a feeling of being together, quite different from a gnostic attitude, which is characterized by emotional distance from its object of knowledge. The latter is based on a *rational* attitude which, in a mode of affect domination, suppresses all affective relations through intentional focusing and structuring so that a simply objectified act, event or existence occurs. The exaggerated anthropological imagination of man as an action-oriented 'brain-machine' makes the contemporary sociological main-stream completely lost when confronted with the fact that feelings and emotions decisively direct human life. In this sociology, the imagination of 'man' is the result of a specific scientism that reduces man to a merely intelligible being – a being who is only accountable to himself. Such cognitive fiction of a subject can be interpreted, from a psychological perspective on science, as a mythical idealization of a being who is *thought* to be controllable, and that is – in consequence – *made* controllable as a scientific object through methodology. Only after the paradigmatic shearing away of feelings, instincts, and affective impulses is it possible to make such an artificially conceived subject appropriate for the construction and maintenance of the (narcissistic) illusion of a world that is epistemologically believed and controllable.[3]

---

2   'dominante Tendenz der kulturellen Entwicklung in der Gegenwart.'

3   The restricting of man to a rationalistic being is the subject of a cultural criticism that is inherent to phenomenology (cf. preface to Schmitz 1997).

I thus assume that the omission of feelings and of the experience of space excludes a specific kind of sociability, which is, however, fundamental for the comprehension of social roles in social systems. To reveal such a methodologically ignored side of human existence, I shall delineate in what follows the basic assumptions of a phenomenology whose scientific object (based on life philosophy) is the result of reflections on pathic situations (in contrast to the construction of a rationalist anthropology in Human Geography). Thereby, I will focus mainly on 'older' approaches in philosophy, psychology, and psychiatry, which during the first half of the twentieth century concentrated on the bodily felt existence of man, which at that time had already been suppressed in the evolution of modern civilization. This tendency, with its sensitivity to the approaches of life philosophy, was mainly represented in the works of Karlfried Count von Dürckheim, Erwin Straus, Ludwig Binswanger, Otto Friedrich Bollnow, Jürg Zutt and others.

This will, however, hardly suffice, as the paradigmatic isolation of human experience from the process of scientific production is not an unintentional (not to say an innocent) lateral consequence of scientific practice. The personal relation between the social role of a professional who is embedded into a system of the official production of knowledge and the self-perception of the life of the same individual through emotions and bodily feelings is too close for such a separation to be maintained without a systematic approach.[4] The decisive question in scientific reflection, therefore, lies in the psychological fact of what makes scientists adhere to such an artificial fiction of human existence, while they live and feel their own body, so that such an attitude obviously and profoundly contradicts all their everyday life experience (cf. also *Psychological consequences for scientific working in the city* in this chapter).

'Lived Space'

Aesthetization processes fill urban spaces with sensitive and emotional qualities and symbols that circulate culturally. Thus their space becomes a site where meaning is linked to a location that forms an *Herumwirklichkeit* [surrounding reality] (according to von Dürckheim 2005 [1932]: 23) which can be experienced personally. Space thus refers to both mental and corporal qualities. In what follows I will focus mainly on the emotional side of experience.

From the perspective of 'lived experience', the character of surrounding can be felt in a corporal-spatial sense as an immaterial 'envelopment' (cf. Frers 2007). In philosophy, especially in phenomenology, these situations are described as 'atmospheres'. Atmospheres can be perceived through 'living corporal

---

4   It is remarkable that such scientific production includes mainly masculine attitudes. Recently, however, reflections on such methodological exclusion of non-rational modes of being have been made mainly by female philosophers, phenomenologists, and psychologists. Between the turn of the twentieth century and the 1920s, however, such discussions were held mostly by male human academics.

communication' (Schmitz 1989 [1978]: 75–109).⁵ At the turn of the nineteenth to
the twentieth century, such corporal communication was understood as *Einfühlung*
[a kind of empathic feeling]. This term was coined under the strong influence
of aesthetical theories with a psychological inclination.⁶ But whereas *Einfühlung*
has been seen predominantly as directed towards the unilateral and passive
component of perception through the perceiving person, the term 'living corporal
communication' actually focuses more on the dialogical character of perception.
Thus the 'partners' of such a communication – things, spatial arrangements,
landscapes – are prone to constant changes because their atmospheric
envelopments are subjected to permanent, sometimes abrupt, modifications so
that the perceiving person is constantly confronted with changing situations. For
example, the mirrored pane of a postmodern skyscraper never appears as a simple
façade, but its appearance is modified permanently due to the sun's course, or to
weather changes, or because of the switching off and on of artificial illumination,
etc. In these situations, the feelings of a perceiving person are always touched, but
in different ways. So the depth of perceptional experience is not only linked to the
(interior) living experience of a person, but also to the depth of external things
that gain their 'life' as shifting physiognomies. Thus, corporal perception exceeds
a simple speech dialogue in which expressions are uttered *expressis verbis*. It is
an event where the surroundings, the *Herum* as von Dürckheim named the 'lived
space', gains its power from sites which address us in a flow of lived events (in
local encounters): 'The field and the wood and the gardens have always been
only a space, and, you, beloved, make them a place' (Goethe: The four seasons).
Thus, in a lived experience, a place greatly exceeds 'the function and the region of
sensual perception' (Binswanger 1947: 16).⁷

So when Count von Dürckheim referred to a *leibhaftiger Herumwirklichkeit*
[reality that surrounds the living body], he was emphasizing less a simply lived
experience as activity, than an experience where space is filled from the outside
with vital qualities. A relationship to the surrounding environment that is 'lived'
in this way is tuned through a process of personal and quasi magic involvement.
Then lived space 'becomes present in its corporal and signifying totality, in each
moment, as an overall positioning, an attitude, a directedness and a mood, which

5   In the recent translation of one of his texts, which has not yet been published,
Hermann Schmitz has authorized the expression 'felt body' for the German word 'Leib'.
However, I personally prefer the translation 'lived body' or even 'living body'. It is thus
possible to put the emphasis on the mutual and active relationship between body and
environment in its openness and presence.

6   Among them the theories of Volkelt (1905ff.), Lipps (1903) and Worringer (1997
[1918]).

7   'die Funktion und das Gebiet der sinnlichen Wahrnehmung.'

one has in his *Innesein* [being inside], in his limbs and his feelings, his body and his heart' (von Dürckheim 2005 [1932]: 26).[8]

The comprehension of such a dialogically lived experience of spatial milieus hardly combines with constructivist approaches which mainly focus on processes of consciousness and rational action. So, for the modern social sciences the full comprehension of human existence (cf. Binswanger 1947: 102) is more than strange, as such a perspective only allows perceiving things as the 'construction' of factual qualities experienced in social processes. Later (in this chapter), it will be shown that this perspective implies ontological limitations, which coincide with a restrictive analysis of those social processes that intentionally create qualities of experience.

The misunderstanding of the difference between the comprehension of the contemporary social sciences of 'lived space' and the proposed 'lived space' by von Dürckheim from the early twentieth century becomes clearer when we turn to Otto Friedrich Bollnow. While von Dürckheim was plainly embedded in classic philosophical phenomenology, Bollnow (also commonly known as a phenomenologist) was influenced thirty years later by modern schools of psychological thought. Thus he interpreted von Dürckheim's expression of 'lived space' simply in a receptive sense. Consequently, his experienced 'lived space' was not a totality of real effectiveness (in German 'Wirklichkeit') but simply an individual psychological experience of an objective 'lived space' (cf. Bollnow 1963: 18f., as well as Hasse 2005: 177ff.). So he cut off the dialogical moment of perception (which almost represents an incorporation) in favour of a receptive understanding grounded in the natural sciences. His 'lived experience' already implies a separation between the experiencing person and the experienced moment, with the experiencing person being located in one place while the experienced moment happens in another. In contrast, von Dürckheim – in his idea of a 'lived space' – (had) insisted on situations of perception where the distance between the 'interior' of a living individual and the 'exterior' situatedness of a given world could be seen as one. Therefore, here the corporal individual was located 'inside' of a lived space in an existential and vital sense: there was no world of things on one side, while a world of experience (of those things) was to be found on the other. Instead, things were always situated and actualized as a self through their nature as an immediate impression, their aesthetic appearance and their atmospheres. From such a phenomenological perspective, a police siren, for example (a common episodic sound element of an urban scene), is not simply a physical and acoustic event of significance, but an atmospheric impression that influences the personal positioning in a mood environment. Erwin Straus has highlighted such situational character by using the example of 'silence': silence can be felt either as eerie, or cosy, or pleasant, or tormenting (cf. Straus 1978 [1930]: 89, Guzzoni 2010).

---

8 'ist in seiner jeweils leibhaftigen und bedeutungsvollen Ganzheit "gegenwärtig" in Gesamteinstellung, Haltung, Gerichtetheit und Zumutesein, man hat ihn im "Innesein", hat ihn in den Gliedern und im Gefühl, in Leib und Herz.'

Our presentation of 'lived space' still needs the complementation of 'lived time', as no modern theory understands space without time. While von Dürckheim was publishing his study on 'lived space', Eugène Minkowski presented his study on the subject of *Le Temps Vécu* [*The Lived Time*] (1933). Minkowski's preliminary remarks on the spatiality of lived experience are very similar to von Dürckheim's way of thinking. He also insists on the conception of a space that is different from Euclidian space, when he points to the fact that life expands into space 'without any proper geometrical extension' (ibid.: 233).[9] Thus, Minkowski interprets 'lived space' as an irrational, non-mathematical and non-geometric space, as an 'extension of the Self into space without spatial mobility'.[10] Such a comprehension of extension does not have a physical, but an emotional, character and oscillates between the poles of corporal narrowness and wideness. Consequently, Minkowski developed his understanding of 'lived space' as a phenomenon of distancing, defining the quality of distance as follows: 'Distance cannot be surmounted; it accompanies us, more joining than separating, neither growing nor diminishing through the distancing from things; it has no limits. To put it bluntly, it has nothing that can be quantified.' (ibid.: 236).[11]

As such, lived distance corresponds to an objectified form of the emotional German 'Weite', the vastness of life. Minkowski refers to 'width' as an ensemble of vital expressions that take place in a space that is before us and organized through lived distances – in its 'width' doing becomes a corporal feeling that is created without involving the power of reason. But even such 'width' needs a primordial production through a potential movement in *factual* (physical) space.

Coming from a background of life philosophy, Rudolph zur Lippe also rejects the conceptual separation of space and time. Using the example of 'lingering' he describes a perception that contemplates its object. 'In lingering we experience an attention which induces us to perceive' (zur Lippe 2010).[12] If we simply followed the logic of those anthropologies that are centred on the theories of cognitive and action theory, lingering can only be understood as a time experienced as passing, a 'zero time', which refrains from any neuro-physiological comprehension of time perception. However, in a phenomenological perspective lingering is a perceptive impression that is produced without any stimulus. No neuro-electrical raw material can be compounded to cognitive structures processed in the Neocortex. To the contrary, the lingering attention is produced by a 'surrounding milieu' in a pathic inclination, similar to mimesis.

Thus, the medium of such a spatial experience of 'lived space' is pathic and not gnostic. Straus had clearly distinguished pathic experience from gnostic experience:

---

9 .   'ohne deshalb eigentliche geometrische Ausdehnung zu haben.'

10   'Ausdehnung des Ichs im Raum ohne Ortsveränderung.'

11   'Elle ne veut et ne peut pas être franchie à proprement parler, puisqu'elle se déplace avec nous, elle relie bien davantage qu'elle ne sépare, elle ne croît ni ne diminue avec l'éloignement des objets, elle n'a pas de limites, elle n'a rien de quantitatif en un mot.'

12   'Im Verweilen gewinnen wir die Aufmerksamkeit, die uns wahrnehmen lässt.'

he interpreted gnostic experience as a rational and conceptive distancing, while the pathic moment of perception which he saw as corporal communication is mimetically involved 'with things because of their sensual givenness'.[13] Such characteristics of things cannot be defined in clear linguistic terms, but appear as situated moods. Consequently, Straus regards the pathic moment as a dimension of original experience. 'For this reason it is nearly inaccessible to any cognitive idea, as these appearances are immediate and presentational, sensual and intuitive, and furthermore pre-communicational.' (Straus 1960 [1930]: 151).[14]

Atmospherical Space

Spaces with vital qualities that can be felt sensitively cannot be analyzed gnostically but are to be perceived in their instant totality as 'atmospheres', 'like a bolt of lightning' according to Schmitz (1998: 33). These atmospheres are evident in a specific manner: we can react to their felt presence, but we are unable to comprehend them through cognitive concepts. Therefore we only relate ourselves to them or live in them, seeking them out when they are attractive, or eluding them if they are cramped or threatening. Such atmospheres are the focus of Gernot Böhme's investigations (he himself referred to H. Schmitz) with examples like dawn, music, church architecture etc. (Böhme 1998).

In urban space, atmospheres are interpenetrating and overlie each other, so that even the concept of 'Urbanity' involves atmospheric aspects. Georg Simmel, in his famous article on *Metropolis and Mental Life* (1957 [1903]), contributed profoundly to our present comprehension of urbanity, which is based on urban citizens who perceive themselves, and are happy to live, in a varied framework unified under the auspices of difference. The metropolis thus develops its cultural potential as a 'place of divergence and attempts to converge' (Simmel 1957 [1903]: 242).[15] Urbanity is not therefore simply the ensemble of material 'urban elements', but a set of urban impressions that are the catalysts of the symbolic and emotional construction of urbanity (like the shiny glamour of refined façades of bourgeois villas, the splendour of expensive department stores, or the economic sublimity of architecturally magnificent buildings). The aestheticized city is therefore in its materiality a medial moment of urbanity, reflecting the heterogeneity and the contradictions of urban life expressed through the characteristics of urban dwellers. According to Simmel, this includes values such as a reserved attitude, bluntness, as well as a lack of concern or interest in one's neighbours (Simmel 1957 [1903]: 232). The emotions that constitute this urbanity are based on the one hand on the mood of personal situatedness, and on the other they are formed through

---

13   'mit den Dingen auf Grund ihrer wechselnden sinnlichen Gegebenheitsweise.'

14   'Es ist darum der begrifflichen Erkenntnis so schwer zugänglich, weil es selbst die unmittelbar-gegenwärtige, sinnlich-anschauliche, noch vorbegriffliche Kommunikation ist, die wir mit den Erscheinungen haben.'

15   'Platz für den Streit und für die Einigungsversuche.'

the aestheticization of the surrounding urban space. Due to the symbolic and emotional complexity of this urbanity, it is difficult to make scientific statements on it, unlike other statements that refer, for example, to transport flows or the forms of buildings in a central metropolitan area. In a wider sense, we even have to assume that urbanity cannot be rationalized at all, which might be one of the reasons for the seemingly endless debates on what urbanity could or should be.

An atmosphere is an emotion with spatial character. According to Hermann Schmitz,[16] 'emotions are not private states of the inner world of the soul, but they are spatially extended atmospheres' (Schmitz 1993: 33) that involve our corporal feelings. Thus, atmospheres are not tri-dimensional, but pre-dimensional and their 'volume' is 'filled' with atmospheres that, like a physical body, have neither surfaces nor divisions; atmospheric volumes do not appear either optically or in tactile forms either. They are essentially experienced through 'power, energy, potentiality and they advance against resistance.' (Schmitz 1988 [1967]: 388).[17] Therefore, if we follow Gernot Böhme, an atmosphere is not a simple *Dort* [being there], but a *Zwischenphänomen* [phenomenon of in-between] (Böhme 2001: 55), an independent form of being that originates from a *Spüren von Anwesenheit* [feeling of presence] (ibid.: 45) located between subjectivity and objectivity.

Schmitz, in contrast to Böhme, understands atmospheres on the objective side but not as objects. For him, the 'place' of an atmosphere is not an *in-between*, but a *being there* that gains its emotive force through feeling as a pre-object. Such *being there* of a potentially perceptive element raises questions about the *being here* of the person who perceives. But even the simple question about the positioning of a viewpoint unsettles the seemingly evident location between a perceiver as *being here* and the perceived object as *being there*. Zutt, for example, puts it this way: I am 'also there', where my view is. 'The reason why I am there, outside, with the things that I can view is that I am not experiencing my view from the dark interior of my body – like the guardian of a lighthouse tower – but that I am in a space which is open to all sides, where my view lingers, and where I am also' (Zutt 1963 [1953]: 348).[18] What can be said in respect to views, can also be referred to atmospheres that subvert the Euclidian dichotomy of here and there. For this purpose, von Dürckheim introduced his term of *Herumwirklichkeit* [surrounding reality].

---

16 My conception of 'atmosphere' mainly refers to Hermann Schmitz, who has developed a profoundly differentiated phenomenology for the analysis of human emotions and feelings, embedding them into ontological positions related to systematic cognitive theories (Schmitz 1964–1980).

17 'Mächtigkeit, Energie, Kraftentfaltung und Andringen gegen Widerstand wesentlich beteiligt.'

18 'Daß ich dort, draußen, bei den erblickten Dingen bin, ist der Grund, warum ich nicht erlebe, aus meinem dunklen Leib hinauszuschauen, wie ein Leuchtturmwärter, sondern in einem allseits offenen Raum zu sein, dort, wo mein Blick ruht, wo ich eben auch bin.'

As atmospheres cannot be directly perceived from a rational distance, they are made perceptible only through pathic participation – something which can be sensitized 'on one's own body, but not as something of one's own body' (Schmitz 1989 [1978]: 118).[19] Consequently, it does not make sense to relate atmospheres to the human body itself. Instead, atmospheres have to be mediated as perceptions via language, gesture, mimicry, music, religion or other forms of expression. What all these forms have in common is that they are founding elements for situations that create individual feelings.

But not every atmosphere affects individual feelings. Atmospheres can also be perceived as collective emotions from a greater distance. Then they are to a certain degree an externality and do not enter into the individual affective-corporal feeling; they remain situations that are cognitively 'controlled' and rationalized. Consequently, Schmitz points to the difference between 'emotion as a perception of feeling an atmosphere, and feeling as an affective being affected' (Schmitz 1993: 48).[20]

This leads us to the difference between the objective and subjective character of atmospheres. Following Schmitz, a subjective atmospheric situation occurs when an individual can make predicates in the first person, whereas an objective situation occurs when everybody is able to perform predicates (ibid.: 51). However, such an objectivity is only possible when the atmospheres exist in a 'pre-objective' sense (Schmitz 1981 [1969]: 103), when they cannot be determined as 'objects' due to there their ontological character. Thus one can end up in atmospheres as in 'weather conditions' (ibid.: 134). When such an atmosphere successfully penetrates the individual in a trickle-down process, it forcefully affects the individual mood. The difference between atmospheres and moods is highlighted by Rainer Maria Rilke in his *Die Aufzeichnungen des Malte Laurids Brigge* [*The Notebook of Malte Laurids Brigge*], where the protagonist is observing, from an emotional distance, the atmosphere of laughing people. The group of people laughing try to integrate him into their affective circle, but Malte feels that he is unable to be integrated: '… and the people stopped me and laughed, and I felt that I should also be laughing, but I could not.' (Rilke 1997 [1910]: 47).[21] His own individual powerful mood prevents him from becoming involved.

In culturally oriented urban research, this proximity between atmospheres and moods deserves special attention (cf. Hasse 2006, 2008). Nearly every staging of urban atmospheres through culture industries is, for example, directed towards the affective moods of individuals. Its aim is to induce certain reactions in the individuals. However, the affective effectiveness of such measures does not generally result in specific determined actions, but is mostly related to emotions

---

19   'am eigenen Leibe, aber nicht als etwas vom eigenen Leib.'

20   'Fühlen als Wahrnehmen des Gefühls als einer Atmosphäre und Fühlen als affektives Betroffensein davon.'

21   '… und die Leute hielten mich auf und lachten, und ich fühlte, dass ich auch lachen sollte, aber ich konnte es nicht.'

predominantly transported unconsciously. As such, urbanity that is produced in urban life develops its own social, political and cultural effects only when penetrated by *affectivity* tainting – as if through incorporation it has tainted both the individual and collective moods.

But an emotional atmosphere can also be produced via the distancing of (cognitive) thinking. In this case, a set of knowledge that is available individually is a pre-condition for statements on emotions. This makes it possible to reflect on one's own affectedness through atmospheric emotions and is accompanied by the ability to criticize. Only when emotions can be *named*, can they also be targeted analytically via criticism. Therefore, most contemporary urban research remains far beyond its own possibilities when investigating the systemic involvement of subjects only on the basis of individuals who master their lives as rational self-determining 'actors'. This is even less the case when emotions, which certainly play an important role in urban life, are not singled out but intermingle via synaesthesia with symbolic meanings. Then, each meaning corresponds to an emotion, and each emotion to a meaning, and as such, life in the city is, like life of the city, less the result of calculated planning than of feelings that originate from urban atmospheres.

Life in cities is permanently changing due to technological, cultural and economic innovations, coinciding with changes in the atmospheres. So, while those atmospheres that Walter Benjamin described in his work on the Passages of Paris (Benjamin 1999 [1989]) were still related to the 'heavy' soil of a local culture grounded in material things, the actual atmospheres of the global lifestyles in our metropolises are increasingly characterized by 'non-material' situations. Consequently, Michel Serres describes our situation as a transition phase from the world of 'Atlas' to the world of 'Hermes': here we liken our work to the work of angels (in the sense of messengers) (Serres 2005). Indeed, the actual advancement of the so-called 'information society', 'knowledge society', or 'service society' has strengthened the immersive forces that have often been overlooked, but which are literally destined to influence human emotions. While public space is progressively disappearing due to the influence of ubiquitously present mass media (cf. Müller and Dröge 2005: 124), the same space becomes available for the 'free' circulation of post-political suggestions for a beautiful life. Such space appeals to feelings that are aroused by fashions and fads that promise non-resistance and non-contradiction when created by the culture industries. To understand these virulent immersive forces, the immaterial media of communication, including atmospheres, have to be investigated in an ontological perspective different from that of things.

Here, Hermann Schmitz's discovery of *Halbdinge* [half-things] reveals a specific ontological class of world-givenness with its own *modus operandi*. Half-things are mediated through 'characteristics' which are more than 'characteristics'. 'They are different from things in two ways: they can disappear and reappear without any need for us to ask where they have been in the mean-time, and they are affectively effective without any responsibility for the causes they exert.' (Schmitz

1994: 80).[22] Schmitz mentions, as examples, sentiments, sounds, smells, velocities and temperatures, but also wind, light, joy, fascination and depression. It is evident that especially under circumstances where high-tech-products are exponentially pluralizing the New Media, these elements are culturally constitutive for the creation of postmodern metropolitan lifestyles. They also influence the forms and rhythms of the formation of identities. These feelings and meanings, which are in a specific manner mediated through half-things are responsible for the immersive hollowing out of the 'austere sense of reality of modernity' (Zimmerli 1988: 17)[23] and so the technological media of immersion are only actualizing what has always been the character of half-things in atmospheres.

Rationality and Irrationality in Urban Life

The diminishing *political* needs for the constitution of a modern self and the complementary advancement of disposing 'lust' for a beautiful life have given aesthetics a new political power and cultural relevance for the formation of identities (Müller and Dröge 2005: 103). Here, emotions have become resources of dissuasion in any systemic process of differentiation on the postmodern dissolution of the frontier between aesthetics and economy. Consequently, the principle of rationality is overlaid by a systemically induced desire that raises more profoundly the question of the relation between rationality and irrationality. Such a new strategic relevance of emotions, especially for the culture industries and ideological strategies, shows that the rationalizing viewpoint of modern social sciences is not sufficient to better understand the 'colonization' of actual urban life worlds (Habermas 1985), as 'rationalism is not an adequate answer to deal with the disorder of cities, and as it ignores our emotional and unconscious reactions to any aspect of urban life' (Müller and Dröge 2005: 170, referring to Kevin Robins).[24] Thus, the question of the relevance of irrationality in urban life is gaining increasing importance.

Today, the term 'irrationality' usually has a negative connotation and is associated with a diminishing level of life-world based knowledge that is blurred by the affects of such irrationality. Now, in late modern life, we encounter the cultural depreciation of all non-rational dimensions, whereas still in the 1920s a researcher like Richard Müller Freienfels assigned a productive relation between the potential of rationality and the irrational element. Irrationality then was understood

---

22   'Sie unterscheiden sich von Dingen auf zwei Weisen: dadurch, daß sie verschwinden und wiederkommen, ohne daß es Sinn hat, zu fragen, wo sie in der Zwischenzeit gewesen sind, und dadurch, daß sie spürbar wirken und betroffen machen, ohne als Ursache hinter dem Einfluß zu stehen, den sie ausüben.'

23   'nüchterne Wirklichkeitssinn der Moderne.'

24   'weil der Rationalismus keine angemessene Antwort ist, mit der Unordnung der Städte umzugehen, und er unsere emotionalen und unbewußten Reaktionen auf jedweden Aspekt städtischen Lebens ignoriert.'

as an attempt to 'assert other means of cognition apart from rational cognition' (Müller-Freienfels 1922: 2).[25] Cognition that was solely based on rationality had been evaluated by him as a one-sided philosophical *Weltanschaung*. Therefore, intellectual cognition and emotional cognition, the latter owing to empathic and creative forces, were put side by side (ibid.: 186f.). For Müller-Freienfels, both forms of cognition were not put into an additive, but into a hierarchical relation of cognitive achievements. 'The *Ratio* has no other function than to generalize what irrationality has produced, in other words it thoughtfully prorogates and expands a process that has started irrationally.' (Müller-Freienfels 1921: 96).[26] Müller-Freienfels interpreted all cognitive knowledge of importance as affectively grounded: 'All great cognitive discoveries are intuitive, which means that they were irrationally constituted and only afterwards became shaped into a rational form.' (ibid.: 96).[27] Though such a position seems to be very distant from the actual hegemonic anthropology that is propagated by the social sciences, it demonstrates its close proximity to phenomenology and its critique of a reduced conception of humane cognitive capabilities limited to abstract forms of thinking.

Paradigmatically, these results have far-reaching consequences for scientific theory. The individual subject can now no longer be described only by cognitive tools that are developed in action theory (for example, by Anthony Giddens), as such theory reduces any perception process to a mere intellectual treatment of data exercised in the material corporality and activity of the human body (cf. Giddens 1984).[28] However, the urban dweller is rooted in 'vital qualities'. He lives in his city in a 'surrounding reality', where the 'surrounding reality' is a 'surrounding space' that is difficult to express.

Such affective dispositions are not simply there; they can be produced methodically from other actors or can arise through an atmospheric *Zeitgeist* which mythically 'charges' places. Such situations can also arise where emotional dispositions are *made* unconsciously, for example, in the case when the stability of a culture is threatened, as Mario Erdheims has demonstrated in his ethno-psychoanalysis (Erdheim 1984: 221): 'The socially unconscious element is so to speak a container which has to receive all perceptions, fantasies, and impulses that could bring the individual subject in opposition to the interests of power' (Erdheim

---

25 'außer dem rationalen Erkennen noch andere Erkenntnismittel zur Geltung zu bringen.'

26 'Die Ratio hat also im Grunde weiter gar keine Funktion, als irrational Entstandenes zu verallgemeinern, d.h. einen irrational begonnenen Prozeß mit Bedacht zu verlängern und zu erweitern.'

27 'Alle großen Erkenntnisse sind intuitiv, d.h. irrational konzipiert und nur nachträglich in rationale Form gegossen worden.'

28 'Perception is organized via anticipatory schemata while simultaneously mentally digesting old' (Giddens 1984: 46).

1988: 275).[29] In his *Heterotopology*, Michel Foucault has focused the subject of such mythically loaded spaces and has demonstrated that the multiple (and mostly rationalized) coding of places is overlaid by a *different* (real) reality through incorporation. Here, the function of 'other spaces' comes into play in order to make a reality disappear through the suggestive realization of utopia (cf. Foucault 2005 [1966]). These mythologies communicate through the architecture of 'other spaces' and are mediated in images, scenes, practices, dispositives. Emotions are implicitly always included. They put a personal *situation* in a specific relation to one's own self, including the surrounding social conditions. Thus, they form a framework of experience that is typical of a specific *Zeitgeist* that offers reasons and motives for specific acts. Today, the culture of great cities is – especially in the guise of a formative economy – saturated with these mythical images that are the bearers of promises. Their dispositional power develops into an 'exterior of language' (Foucault 1987 [1966]: 51).[30] As symbols, they are gradually transformed synaesthetically into emotions to finally incorporate meanings of systemic importance in the subjects. Three examples will elucidate these situations.

Urban Illumination – Sentimental Lighting

Light is one example for the exteriority of an atmospheric lived space. In Christian mythology it appears as a *theophore Metapher schlechthin* [divine metaphor of its own] (Thiel 2006: 233). Here, the allegory of 'divine light' provides the field of sensual perception with religious meaning. Thus, all those monumental light events that are organized at postmodern spectacular cultural events are intended to create a secular atmosphere of experience which presents the city as beautiful and capable of enthusiasm.

In this context, all the Christmas illuminations that return to our cities every year attempt to create an emotionalized situation in the (consumer-)space of the city. The arrangers of this sentimental lighting rely on a huge reservoir of conveyed knowledge when developing their aesthetical oeuvre. The sustained effect of such commonly-used methods is due to synaesthetical effects, which are not necessarily intended, but which are the result of the application of intuitive ('irrational') knowledge.[31] The atmosphere of Christmas cannot be decoded semiotically in everyday life, step by step, but is sensitized in specific situations as a totality and

---

29   'Das gesellschaftliche Unbewusste ist gleichsam ein Behälter, der all die Wahrnehmungen, Phantasien, Triebimpulse aufnehmen muß, die das Individuum in Opposition zu den Interessen der Herrschaft bringen könnte.'

30   For Foucault, the exterior is an abyss beyond the rationalistic orders of thinking and experience – not as a dark crevice of non-reason but with its own rationality which put reason in its specific and limited place.

31   Zur Lippe (1987) describes such knowledge as 'sensual consciousness'. It is characterized by a high degree of incorporation of knowledge.

then its emotional effects and meanings are simply incorporated. The condition for such sentimentalized atmospheres is one's affective readiness to encounter the background of Christian Mythology in a Christmas atmosphere. However, the evocated atmosphere can cause such a strong sentimental maelstrom that even secular scenes of urban illumination are tainted and transformed to Christmas atmospheres, at least if the affected areas offer possibilities for an integration into the logic of Christmas. Consequently, the illuminated sandstone façade of a prestigious historicist building, for example, is compatible in a sensual and symbolic sense with the *soft* atmospheric image of a Christmas illumination when it combines the mild lighting that covers the raw yellowish sandstone with the symbolism of Christmas, uniting the sensual 'warmth' with the symbolic 'warmth' of Christian-mythological significance. Based on the ubiquitous character of knowledge about the meaning of light in Christian ideology, the atmospheres that have been staged in this respect develop a powerful suggestive force, even if there is a psychological reserve that can secure an affective distance towards Christian Mythology. In such a situation, the atmosphere is active and its tuning effect can be fully perceived (cf. the description of Rilke).

Gardens – Emotional Green

Another example of the atmospheric construction of lived experience is urban green. The methodical organization of urban green spaces is legitimized in modernity in a pragmatic way. Though the function of green belts, gardens and parks is certainly justified for reasons of their specific micro-climate, they are primarily welcomed as contemplative spaces for recreation, relaxation and excitement. Thus, they fulfil a psychological stress relief function in a socially and physically condensed space of the city. In contrast to such space, the vast landscape gardens of the eighteenth and nineteenth centuries (on the edge of the cities) were immersive spaces of symbols and emotions created as 'planted' atmospheres to evoke moods, to express romantic images of nature and to be placed into the symbolic framework of illuminist political ideas. The conceptual basis for such emotionalized landscaping was provided by Christian Laurenz Hirschfeld (1779–1785) and his comprehensive theory of landscape art, in which he developed detailed instructions for a garden design that enabled the induction of certain atmospheres to harmonize sensuality and reason (cf. Niedermeyer 1996: 54).

A specific garden which until today is associated with the transmission of emotions and has left profound vestiges in the physiognomy of the European city is the Christian cemetery. Today, cemeteries present themselves as a colourful mosaic of different forms. The modifications that these European cemeteries underwent at the turn from the eighteenth to the nineteenth century are very instructive. At that time their symbolic forms changed to those of the English Garden (and thus to 'nature'). Here, the romantic perception of nature not only modifies the *cognitive* imagination of death, but also the *emotional* relation to the finiteness of life. Thus, the general trend to the idealization of nature has led coincidentally to a symbolic 'smoothening' of the concept of death via the spatial and material conception of the cemetery. Though Hirschfeld himself had little influence on the architecture of cemeteries, his remarks exemplify how the programming of cemetery space could and should be implanted through numinous atmospheres. The burial sites, which 'originally had to be placed outside of the cities', should be 'quiet, secluded and serious ... They belong to the melancholic type of gardens. Such a place, however, has to be preserved with low walls, or ditches, or fences, but not with an enclosure of fear. ... A dark pine forest in the vicinity, a dull murmuring of falling water close by intensifies the holy melancholy of the place. Its trees are destined to announce the mourning scene through their brown and dark leaves.' (Hirschfeld 1785: 118).[32] Such instructions clearly reveal how Hirschfeld had used the synaesthetical potential of the staging of atmospheres to educate for a methodical sensual communication based on the intentional effects of specific moods. He wanted to create a counterpart to the 'noisy stage of the world' to respond adequately to the importance of such a place: 'This totality has to represent a great, earnest and solemn painting that has nothing to do with a dreadful and horrible scene, but which stirs up the imaginative forces, and which coincidentally moves the heart to merciful, tender and soft melancholic feelings.' (Hirschfeld 1785: 119).[33]

Architecture – Presenting Myths

A third example of atmospheric experience is architecture. When architecture stirs people's feelings, it is usually an argument about whether buildings are suitable or unsuitable, beautiful or ugly, successful or a failure. However, such quarrels about forms, rules and judgements of taste are the expression of deeper-rooted feelings, which cannot usually be articulated in the debate in a concrete form. Such a confused link between rationalization and irrationality becomes clearer in German semantics, for example, in the case of the connotative relation between *Mut* and *Gemüt*: In German, the expression *Mut* means a feeling of courage.

---

32 'Begräbnisplätze, die demnächst außer den Städten anzuweisen sind, müssen eine Lage haben, die reinigenden Winden den Zugang verstattet, und eine ruhige, einsame und ernste Gegend. Sie gehören zu der melancholischen Gattung von Gärten. Der Platz muß allerdings durch niedrige Mauern, oder Graben, oder Zaun eine Beschützung, aber keine ängstliche Einsperrung haben. ... Ein finsterer angränzender Tannenwald, ein dumpfigtes Gemurmel fallender Wasser in der Nähe, vermehrt die heilige Melancholie des Orts. Die Bäume müssen durch braunes und dunkles Laub die Trauer der Scenen ankündigen.'

33 'Das Ganze muß ein großes, ernstes, düsteres und feyerliches Gemälde darstellen, das nichts Schauerhaftes, nichts Schreckliches hat, aber doch die Einbildungskraft erschüttert, und zugleich das Herz in eine Bewegung von mitleidigen, zärtlichen und sanftmelancholischen Gefühlen versetzt.'

This is etymologically linked to *Gemüt*, a type of mood that can be seen as an emotional expression. Therefore, the German *Erregung der Gemüter* [the stirring of people's feelings] serves as an emotional base for (rational) judgements of validation, bridging the gap between rationality and emotion. Such a relation applies, for example, in the case of those 'prestigious' architectural buildings that do not simply fulfil a pragmatic functional 'sense', but also serve representational functions (with a symbolic meaning) relating the building to 'something else'; these may be affects such as winning affection, pleasure or affirmation, so that symbolical and emotional functions are also included in these constructions.

However, it would be naive to measure the 'appeal' of architecture only by its degree of aesthetical pleasure. Aesthetical constructions are embedded in the foundation of a *lived* relation to 'something'. Here, architecture serves as a tool for bridging the distance between its profane tasks on the one hand and its cultural function of symbolic orientation on the other. The building of a prison, for example, has a pragmatic function as an enclosure and disciplining place for delinquents (as a hermetic architecture). To fulfil such a task, for example when an insurmountable wall is needed, it would be sufficient to use suitable building materials. However, when the entrance buildings of prisons appear sumptuous beyond the mere engineering necessities, they represent magnitude and gravity (as was the case in the nineteenth century), and such aesthetics communicate the feeling of dignity and honour. This is appropriate for the authority of the state and the power of its institutions. Thus, the meaning of 'punishment for the benefit of the society' receives an additional degree of credibility. If the state had simply announced the importance of its prisons, it would have been reduced to the limited forces of its words and abstract symbolizations. However, when state institutions are integrated into an emotional humane regime through the power of allegories, their specific sustainability is reinforced through the symbolic appeal.

Such a connection between emotion and technical function is more much hidden in profane functional buildings. These seem to be very distant from any symbolism. For example, multi-storey car parks usually have a very evident functional significance. They bring together the parked vehicles. Without the spatial and temporary agglomeration in particular places, the city traffic would simply collapse. Thus, the idea of a car park at the beginning of the twentieth century in Europe (a development that occurred some decades earlier in the USA) seemed to have solely a functional purpose, and it appears to be pointless to look for the aesthetical side of car park buildings, either in their interior or their façades. However, the first car parks in Europe were temples that enshrined the automobile (cf. Hasse 2007). Beyond their silo-like storing function they also positioned the automobile culturally as a symbolic means for announcing progress in a new era. But the aesthetic force of these first buildings of a new type, usually designed by prestigious architects, did not simply produce a cultural surplus of meaning, but represented a symbolic link between culture and economy.

Surprisingly, when the aesthetics of multi-storey car parks passed through different historical periods they were less linked to the improvement of their pragmatic (especially technical) functions than to symbolic-cultural orientations that integrate the automobile into society. In this context, aesthetics have always worked as an appeal for emotions – oscillating between programmatic ugliness (in times of ecological proscription against the automobile) and postmodern glamour with its post-critical fetishism (in present times). Since the absurd increase of commuting, a third function has to be added: the contra-factual suggestion of a better life-style, one that is based on a transport technology which enables culturally and politically unlimited private transport. Under these circumstances, the dissuasive function of architectural aesthetics reveals its benefits for the automobile industry. Now, in postmodern times, aesthetics and economics melt in one indivisible functional relation.

All three examples that have been highlighted reveal the importance of emotions for the cultural constitution of society. The feelings involved serve the auto-poetic optimization of systemic 'meaning' and are able to develop a specific and effective sustainability when isolated against individual and collective processes of rational understanding, which means after being 'made unconscious' (Erdheim 1984, 1988). In this context, the culture industries appeal basically not to reason, but to the emotions.

Psychological Consequences for Scientific Working in the City

It has been demonstrated that feelings and emotions are not only an important element in human life, but they are also incorporated into systemic relations (especially of an economic, cultural and political nature) when used for pragmatic interests. In the actual process of civilization they are pushed away into the fields of privacy, religion, psycho-sociology and therapy, a fact which becomes especially visible in the paradigmatic construction of German-speaking social sciences, which is currently greatly influenced by Anglophone debates. Thus the renewed importance of the neurosciences reinterprets feelings as evaluative cognitions (cf. Roth 1994), and philosophical research on emotions has undergone a certain 'rationalization' when describing feelings as basically oriented towards objects (cf. Weber-Guskar 2009: 35f.). However, emotional moods are not directed. They are characterized by 'thematic' diffusion, so that they cannot be adequately treated from this perspective. What is more important, however, is a specific trait of the body that becomes perceptible only in the German-speaking debate, where it is possible to differentiate between the 'body' as a mere physical corpus – derived from a Latin tradition based on the idea of 'corpus' -, and *Leib*, a lived body that is etymologically linked to the Anglo-Germanic word 'life'. Actually, the strong influence of the Anglophone and French debates has expelled all combined corporal-mental elements, like emotions, from the actual paradigmatic apparatus for the analysis of the social world, pushing these elements into the field of mere 'mental' facts. Earlier on, the German debate treated these elements (especially in phenomenological approaches) under the category of *Leib*. Such an approach of rationalization has consequences, as can be seen in the critics of Euclidean Space.

Zur Lippe interprets the actual Euclidization of space as a *Geographie der Gewalt* [geography of violence] that abstracts the bodily felt emotional space (including atmospheres and moods) from the reality of the given world (zur Lippe 2010). This is the result of a biased comprehension of Man where Man is seen, within an action-oriented model, as someone who has power over himself. In *Le Différand* [*The Differend*] (1988 [1983]), Lyotard branded such submission of alterity under the pressure of a unifying formula as an illegitimate performance of violence. But the hegemonic consensus of the *scientific community* strengthens such 'geography of violence', as their approaches separate rational thinking from emotional feeling into a theoretical configuration of an anthropology which is simply based on pure rationality, avoiding any 'irrationality' in its methods.

So the exclusion of human emotions and the 'purification' of any irrationality are not simply a problem of theory, but they are fully applied in empirical research. Though irrationality can be simply seen as the other side of rationality, mainstream social theory only communicates commensurable segments of a 'person'. Consequently, a fiction of Man is *construed* in the name of constructivism that complies with a cognitive model of a calculable individual.

In this sense, the *Geometrisierung des Menschen* [geometrization of man] (zur Lippe 2010) profoundly questions science from a psychological standpoint. So what had already bothered the bourgeois society of the occidental world (cf. zur Lippe 1996: 101), actually becomes a precarious idiosyncrasy for scientific discourses and thinking: the living body of Man and his emotions are understood as 'irrational', and such a cognitive fixation prevents scientists from understanding that it is also feelings that are a fundamental in any scientific research practice – not least in the (silent) consensus that connects the scientific community as a social link. In this consensus, the discursive acceptance of a set of validated theories leads to commonly accepted modes of thinking. Consequently, Carola Meier-Seethaler (following Ludwig Fleck) points to the fact that *Denkstile* [modes of thinking] can never be validated as 'pure' in a rationalist or epistemological sense, but are the result of hegemonic *Denkstimmungen* [moods of thinking] (Meier-Seethaler 2007a: 79). This means that the supposed argumentative consensus on theoretical premises also includes an affective dimension. What is *cognitively* approved is also embodied in an emotional consent. Only if this requirement is fulfilled, can a mode of thinking be approved as 'correct' in science. This hidden emotional character of scientific discourse encourages the articulation of scientific criticism in a sustainable way. It is the will to criticize – coinciding 'with a mood, a passion, a feeling, moral or ethics that is linked to critique' (Demirovic 2008: 10)[34] – which precedes any rational statement. When the Norwegian philosopher Arne Næss expresses his 'reluctance to accept that there is no room for feelings in

---

34   'der mit einer Stimmung, einer Leidenschaft, einem Gefühl, mit einer Moral oder Ethik der Kritik verbunden ist.'

intellectuals' (Næss 2009: 24),[35] a statement which he makes when justifying his life-philosophical research on the relation between emotion and aesthetics, then he is positioning his critique not beyond science, but is emphasizing the importance of those cognitive forces. But these are marked by the hegemonic *Zeitgeist* of the so-called 'scientific community' as extra-scientific. However, it is specifically these forces that induce the creative initialization and the progressive continuation of research processes in their broader sense.

Therefore, interaction between affects and cognitions is not a 'threat' for the consensus of science, but an expression of a relation between emotion and mind that is inseparable from an anthropological perspective. Therefore, only an exaggerated perspective on rational *scientific* thought requires 'purity' and, consequently, denounces the mingling of cognition and affects as an unacceptable 'impurity'. In general, however, each case of a commonly shared 'mood of thinking' creates an emotional framework for 'affective logics' (Ciompi 1988), so that arguments are only the cognitive clothing of an *Denkkollektiv* [assembly of thinkers] (Meier-Seethaler 2007b: 37). Instead, according to Hilge Landweer, requiring pure rationality even encompasses the danger of *kollektiver Selbsttäuschung* [collective self-illusion] (Landweer 2007: 65). In the same way, Meier-Seethaler also emphasizes the implicit threat of irrationality in the centre of those cultures of thought that present themselves as extremely rational while hiding and even denying any emotional personal involvement in the research process. 'This means that a society which hardly accepts a culture of emotion is especially prone to irrational undercurrents on the collective level.' (Meier-Seethaler 2007b: 37).[36] Such affective 'unmaking of consciousness' in the research process, created by excluding any affective element in theory is socialized among the younger generation of scientists through scientific traditions, for example, when university students are obliged, as is logically consistent, to refrain from using the first person in scientific papers and presentations.

What has been coined in education as a 'hidden curriculum' (Jackson 1968) or 'poisonous pedagogies' (Rutschky 1977) and what has now become a target in the crosshairs of a philosophical critique of pedagogy, is still preserved in the social system of science. Such a 'black socialization' causes the exclusion of any ethical impetus for the individual, preventing him from reflecting on his own responsibility for the production of his creations. Therefore, in the social sciences and humanities, the fear of the dangers of irrationalism has become irrational in itself. About seventy years ago, Willy Hellpach (1937) demonstrated the irreducible character of the irrational grounding of scientific research from a psychological point of view. Such a condition of research could be deduced from the fact that Man only has a limited capacity to refrain from everything that befalls him, or

---

35   'Widerwillen bei der Art von Zuweisung, die den Intellektuellen keinen Raum für Gefühle zugesteht.'

36   'Auf der kollektiven Ebene heißt dies, dass eine Gesellschaft, die wenig Wert auf eine bewusste Gefühlskultur legt, für irrationale Unterströmungen besonders anfällig ist.'

gains significance when he becomes aware of his situation. It is the 'basic character of individuation that allows only limited rational generalizations and that cannot directly approach the essentials' (Hellpach 1937: 21).[37] '... even "perception" can hardly be separated from experiencing, aspiring, longing; without desired images there would not be any passion for research, and without that passion there would be no scientific result' (ibid.: 53).[38]

However, the ignition of new ideas through extra-rational impulses (even if they are not revolutionary) is protected against a permanent irrationalism, because science systematically and constantly makes its irrationality accessible to processes of rationalization. Hellpach characterized such a situation as an *irrationale Ursprungssituation* [irrational origination], positioning it as a link to formal rationalization. A very different psychological aspect is the irrational content of life-worlds, even including *Widersinn* [absurdities] (Hellpach 1937: 21). In the hegemonic tendencies of the social sciences such aspects are usually ruled out through the theoretical exclusion of emotions of their own. Even a simple inclusion of the knowledge of the researcher's life-world experience contrasts categorically with the practice of abstraction in scientificism. The avoidance of a discussion on such a structural contradiction is thus inherent to the system. Consequently, the collective ascription of dignity immunizes the Ego against the claims of scientific comprehension, and both the individual and the scientific collective are mutually protected against each other in their respective identities. But Hellpach's suggestion that all statements include an irrational content touches the Achilles' heel of scientific self-identity, as scientific identity is based on the illusion 'that science is only a making-aware of an underlying reality, and not a production of a position of the Ego in relation to reality in its totality.' (Müller-Freienfels 1936: 72).[39]

As such, the affective undercurrents of rationality in scientific practice are systematically integrated both on the formal and the content level, and they are auto-poetically generated throughout the scientific process. The selection and fixed attention on one or a few theoretical corpuses among all the theoretically available theories must be evaluated as being produced through affective orientation. The discursive practices of power in specific disciplines, which are usually communicated in a proto-rational manner, also direct the scientist's thinking in a way that he or she refers directly to those theories where an emotional identification is libidinously construed. Under such a condition, an informal group of researchers

---

37 'Wesen der Individuation, dass sie rationaler Generalisation nur begrenzt und nicht in Anwendung ihres Wesentlichen zugänglich ist.'

38 'auch das "Wahrnehmen" löst sich nur unvollkommen vom Erleben, vom Wünschen, von der Sehnsucht; ohne Ergebniswunschbilder gäbe es überhaupt keine Forschungsleidenschaft, und ohne sie keine wissenschaftliche Erkenntnis.'

39 'daß es in der Wissenschaft nur auf ein Bewusstmachen der Wirklichkeit ankommt, statt – wie es in der Tat ist – auf eine ganzheitliche Stellungnahme unseres Ich zur Wirklichkeit.'

can also be interpreted as an affective *community*. Then, but only then, can this group be recognized socially in a form which excludes the irreducible dimensions of human existence.

Actually, the city – as the predominant modern life world environment – appears in a basic opposition between rationality and irrationality. This binary contradiction has become vital for individual and social life. While science is evidently obliged to rationally distance itself from its object, favouring the rational side, such a basic paradigm does not mean that processes of emotional experience should be withdrawn from any analytical investigation. Instead, they too are an inseparable dimension of the experienced reality – apart from the objectified reality of the city's elements such as buildings, streets, squares and lakes, wealth and poverty. Here, the facets of subjectivity that have been discussed in this chapter are expressed simultaneously on different levels: from a life world perspective, the city is an emotionally lived space influenced by the moods of fascination and idiosyncrasies. However, such mediums of experience are not simply there – for example, like weather, day and night, summer and winter. They appear in the condensed space of the city in the form of symbolic, economic, technological or other framings, following tendencies for hermetic closing within their respective systems, but induce individuals to experience and act in a specific way. Such a situation puts the epistemological focus not only on social actors who live in the city, in pathic attitudes, with their own performative practices, but also on professional actors who are inclined to manipulate the life of these habitants rationally. From this perspective, an open space appears in urban geography where the complex and contradictory lived space is construed and constituted between rationality and irrationalty.

References

Abramson, P.R. and Inglehart, R. 1995. *Value Change in Global Perspective*. Ann Arbor: University of Michigan Press.

Baudrillard, J. 1983. *Laßt euch nicht verführen!* Berlin: Merve.

Benjamin, W. 1999 [1989]. *The Arcades Project*. Cambridge, MA, London: Belknap Press.

Binswanger, L. 1947. *Ausgewählte Vorträge und Aufsätze, vol. 1. Zur phänomenologischen Anthropologie*. Bern: Francke.

Böhme, G. 1998. *Anmutungen. Über das Atmosphärische*. Ostfildern: Ed. Tertium.

——— 2001. *Aisthetik. Vorlesungen über Ästhetik als allgemeine Wahrnehmungslehre*. München: Fink.

Bollnow, O.F. 1963. *Mensch und Raum*. Stuttgart: Kohlhammer.

Ciompi, L. 1988. *The Psyche and Schizophrenia: The Bond between Affect and Logic*. Cambridge, MA: Harvard University Press.

Demirovic, A. 2008. 'Leidenschaft und Wahrheit. Für einen neuen Modus der Kritik', in *Kritik und Materialität*, edited by A. Demirovic. Münster: Westfälisches Dampfboot, 9–40.

Erdheim, M. 1984. *Die gesellschaftliche Produktion von Unbewußtheit*. Frankfurt am Main: Suhrkamp.

———— 1988. 'Der Alltag und das gesellschaftlich Unbewußte', in *Psychoanalyse und Unbewußtheit in der Kultur*, edited by M. Erdheim. Frankfurt am Main: Suhrkamp, 269–78.

Foucault, M. 1978 [1977]. 'Die Machtverhältnisse durchziehen das Körperinnere. Ein Gespräch mit Lucette Finas', in *Dispositive der Macht. Über Sexualität, Wissen und Wahrheit*. Berlin: Merve, 104–17.

———— 1987 [1966]. 'Das Denken des Außen', in *Von der Subversion des Wissens*. Frankfurt am Main: Fischer Taschenbuch, 46–68.

———— 2005 [1966]. 'Die Heterotopien', in *Die Heterotopien. Der utopische Körper. Zwei Radiovorträge mit einem Nachwort von Daniel Defert*. Frankfurt am Main: Suhrkamp, 7–22.

Frers, L. 2007. *Einhüllende Materialitäten. Eine Phänomenologie des Wahrnehmens und Handelns an Bahnhöfen und Fährterminals*. Bielefeld: Transcript.

Giddens, A. 1984. *The Constitution of Society. Outline of the Theory of Structuration*. Oxford, Cambridge: Polity Press.

Guzzoni, U. 2010. 'Nächtliche Geräusche. Raumerfahrungen in literarischen Bildern', in *5. Jahrbuch Lebensphilosophie*, edited by J. Hasse and R. Kozljanič. München: Albunea, 83–95.

Habermas, J. 1985. *Die neue Unübersichtlichkeit*. Frankfurt am Main: Suhrkamp.

Hasse, J. 2005. *Fundsachen der Sinne. Eine phänomenologische Revision alltäglichen Erlebens*. Neue Phänomenologie, vol. 4. Freiburg, München: Karl Alber Verlag.

———— 2006. 'Atmosfere e tonalità emotive. I sentimenti come mezzi di comunicazione'. *Rivista di Estetica*, 33(3), 95–115.

———— 2007. *Übersehene Räume. Zur Kulturgeschichte und Heterotopologie des Parkhauses*. Bielefeld: Transcript.

———— 2008. 'Die Stadt als Raum der Atmosphären. Zur Differenzierung von Atmosphären und Stimmungen'. *Die alte Stadt*, 35(2), 103–16.

Hellpach, W. 1937. *Schöpferische Unvernunft? Rolle und Grenze des Irrationalen in der Wissenschaft*. Leipzig: Meiner.

Hirschfeld, C.C.L. 1779–1785. *Theorie der Gartenkunst, vol. 1–5*. Leipzig: M.G. Weidmanns Erben und Reich.

———— 1785. *Theorie der Gartenkunst, vol. 5*. Leipzig: M.G. Weidmanns Erben und Reich.

Jackson, P.W. 1968. *Life in classrooms*. New York, London: Holt, Rinehart & Winston.

Kluck, S. 2008. *Der Zeitgeist als Situation*. Rostocker Phänomenologische Manuskripte, vol. 3. Rostock: Universität Rostock.

Landweer, H. 2007. 'Sozialität und Echtheit der Gefühle. Die Ethik des Echten, soziale Normen und Imagination', in *Kritik der Gefühle: Feministische Positionen*, edited by A. Neumayr. Wien: Milena, 63–91.

Lipps, T. 1903. *Grundlegung der Ästhetik, vol. 1*. Leipzig, Hamburg: Voss.

Lyotard, J.-F. 1988 [1983]. *The Differend. Phrases in Dispute.* Minneapolis: University of Minnesota Press.

———— 1989. 'Das Interesse des Erhabenen', in *Das Erhabene. Zwischen Grenzerfahrung und Größenwahn*, edited by C. Pries. Weinheim: VCH Acta humaniora, 91–118.

Meier-Seethaler, C. 2007a. *Macht und Moral*. Zürich: Xanthippe.

———— 2007b. 'Gefühl und Urteilskraft. Emotionale Vernunft aus feministischer Sicht', in *Kritik der Gefühle: Feministische Positionen*, edited by A. Neumayr. Wien: Milena, 30–44.

Minkowski, E. 1933. *Le Temps Vécu. Etudes Phénoménologiques et Psychopathologiques*. Collection d'Evolution Psychiatrique. Paris: d'Artrey.

Müller, M. and Dröge, F. 2005. *Die ausgestellte Stadt. Zur Differenz von Ort und Raum*. Bauwelt Fundamente, vol. 133. Basel, Berlin: Birkhäuser.

Müller-Freienfels, R. 1921. *Philosophie der Individualität*. Leipzig: Meiner.

———— 1922. *Irrationalismus: Umrisse einer Erkenntnislehre*. Leipzig: Meiner.

———— 1936. *Psychologie der Wissenschaft*. Leipzig: Barth.

Næss, A. 2009. 'Es gibt keine reine Ästhetik ohne Gefühl. Interview mit Henrik B. Tschudi'. *Information Philosophie*, 37(2), 24–29.

Niedermeyer, M. 1996. 'Aufklärung im Gartenreich Dessau-Wörlitz', in *Weltbild Wörlitz. Entwurf einer Kulturlandschaft*, edited by F.-A. Bechtholdt and T. Weiss. Ostfildern: Hatje, 51–65.

Rilke, R.M. 1997 [1910]. *Die Aufzeichnungen des Malte Laurids Brigge*. München: Deutscher Taschenbuch Verlag.

Roth, G. 1994. 'Verstand und Gefühle'. *Kunstforum International*, 22(126), 118–26.

Rutschky, K. 1977. *Schwarze Pädagogik. Quellen zur Naturgeschichte der bürgerlichen Erziehung*. Frankfurt am Main: Ullstein.

Schmitz, H. 1964–1980. *System der Philosophie, 5 Bände in 10 Bänden*. Bonn: Bouvier.

———— 1981 [1969]. *System der Philosophie, vol. 3. Der Raum. Teil 2: Der Gefühlsraum*. Bonn: Bouvier.

———— 1988 [1967]. *System der Philosophie, vol. 3. Der Raum. Teil 1: Der leibliche Raum*. Bonn: Bouvier.

———— 1989 [1978]. *System der Philosophie, vol. 3. Der Raum. Teil 5: Die Wahrnehmung*. Bonn: Bouvier.

———— 1993. 'Gefühle als Atmosphären und das affektive Betroffensein von ihnen', in *Zur Philosophie der Gefühle*, edited by H. Fink-Eitel and G. Lohmann. Frankfurt am Main: Suhrkamp, 33–56.

———— 1994. *Neue Grundlagen der Erkenntnistheorie*. Bonn: Bouvier.

————— 1997. *Höhlengänge. Über die gegenwärtige Aufgabe der Philosophie*. Berlin: Akademischer Verlag.

————— 1998. *Der Leib, der Raum und die Gefühle*. Ostfildern: Ed. Tertium.

Serres, M. 2005. *Atlas*. Berlin: Merve.

Simmel, G. 1957 [1903]. 'Die Großstädte und das Geistesleben', in *Brücke und Tür. Essays des Philosophen zur Geschichte, Religion, Kunst und Gesellschaft*. Stuttgart: Koehler, 227–42.

Straus, E. 1960 [1930]. 'Die Formen des Räumlichen. Ihre Bedeutung für die Motorik und die Wahrnehmung', in *Psychologie der menschlichen Welt. Gesammelte Schriften*. Berlin: Springer, 141–78.

————— 1978 [1930]. *Geschehnis und Erlebnis. Zugleich eine historiologische Deutung des psychischen Traumas und der Renten-Neurose*. Berlin: Springer.

Thiel, R. 2006. 'Stichwort "Licht"', in *Metzler Lexikon Ästhetik*, edited by A. Trebeß. Stuttgart, Weimar: Metzler, 233–34.

von Dürckheim, K. 2005 [1932]. 'Untersuchungen zum gelebten Raum', in *Graf Karlfried von Dürckheim: Untersuchungen zum gelebten Raum*. Natur – Raum – Gesellschaft, vol. 4, edited by J. Hasse. Frankfurt am Main: Institut für Didaktik der Geographie, 11–108.

Volkelt, J. 1905ff. *System der Ästhetik, vol. 1. Grundlegung der Ästhetik*. München: Beck.

Weber-Guskar, E. 2009. *Die Klarheit der Gefühle. Was es heißt, Emotionen zu verstehen*. Berlin, New York: de Gruyter.

Welsch, W. 1993. 'Das Ästhetische. Eine Schlüsselkategorie unserer Zeit?', in *Die Aktualität des Ästhetischen*. München: Fink, 13–47.

Worringer, W. 1997 [1918]. *Abstraction and Empathy: A contribution to the psychology of style*. Chicago: Ivan R. Dee.

Zimmerli, W.C. 1988. 'Das antiplatonische Experiment', in *Technologisches Zeitalter oder Postmoderne?* München: Fink, 13–35.

zur Lippe, R. 1987. *Sinnenbewußtsein*. Reinbek: Rowohlt.

————— 1996. 'Karlfried Graf von Dürckheim. Einige Züge seiner Wirkung und Bedeutung', in *Der Mensch als Zeuge des Unendlichen. Karlfried Graf von Dürckheim zum 100. Geburtstag*, edited by T. Arzt, M. Hippius-Gräfin von Dürckheim and J. Robrecht. Schaffhausen: Novalis, 87–107.

————— 2010. 'Zeit – Ort im post-euklidischen Zeitalter', in *5. Jahrbuch Lebensphilosophie*, edited by J. Hasse and R. Kozljanič. München: Albunea, 109–20.

Zutt, J. 1963 [1953]. '"Außersichsein" und "auf sich selbst Zurückblicken" als Ausnahmezustand', in *Auf dem Wege zu einer anthropologischen Psychiatrie. Gesammelte Aufsätze*. Berlin, Göttingen, Heidelberg: Springer, 342–52.

Chapter 5
Aesthetics and Design:
Perceptions in the Postmodern Periphery

Ludger Basten

Introduction

Michael Dear, Edward Soja, Mark Gottdiener and many others have provided us with wide-sweeping and far-reaching descriptions and interpretations of cities in postmodern times (e.g. Dear 2000, Gottdiener 1994, Soja 1989, 2000). They have spoken and written of the multitude of processes that are reconfiguring urban spaces today, of the bits and pieces – both material places and social practices – that are to be found in such a postmodern urbanism. And they have tended to portray Los Angeles as the place where 'it all comes together' (Soja 1989: 190ff.).

These writings, both in their collectivity (to avoid the rather inappropriate term 'totality' in this context) but also in their individual variety, are immensely rich in theoretical insight, empirical observations and theory-driven interpretations. Accordingly, these authors have profoundly influenced or even changed the way we see and understand cities or the 'urban' in our present times. They have also changed the way we interpret and possibly even experience urban life as such. They have provided new perspectives for us, as urban dwellers, to find and construct meaning in urban places and in our daily lives lived within them.

Of course, the writings and ideas of the aforementioned authors are not universally accepted as a kind of newly dominant or even hegemonic interpretation of the urban or urban theory. Postmodernism or postmodern urbanism – both as an analytical set of theoretical ideas and, arguably even more so, as a normative ideology – has antagonized many and has been fiercely criticized right from the beginning, when writers such as Soja started to frame their ideas using the postmodernism label. The thrust of such criticisms has been varied. Some voices have renounced postmodernism for its alleged ideological or political stance (interestingly enough, such criticism has come from the left as well as from the right of the political spectrum). Others have focused on theoretical issues, both ontological or epistemological in character. Adherents of modernist theories of knowledge – 'traditional' or 'Marxist' – find it hard to accept a kind of metatheory that purports not to privilege particular ways of knowing or of *constructing* truths. Rather, they would maintain that there exist very specific ways of *discovering* facts and truths (i.e. 'scientific' methodologies) and thus reassert the premise that there is one, and only one (!), reality to be perceived, discovered, understood and explained (e.g. Himmelfarb 1997, Harvey 1989). Thus the pluralistic viewpoints and stances of postmodernism have not found much favour in such quarters.

Furthermore, like other academic theories, perspectives and topics, the theories of postmodernism and postmodern urbanism also seem to have gone through lifecycles with their own phases of being 'en vogue' and phases of being out of fashion. So in many countries and in many disciplinary or multi-disciplinary academic discourses postmodernism is now largely considered passé – at least as a set of academic theories (Becker 1997: 2). In some discursive formations, German geography being one, I would argue, it never truly arrived in the first place. While some select few picked up on these ideas and went on to discuss their theoretical merits, postmodernism, by and large, was simply ignored or lightly brushed aside. Rather, other theoretical buzzwords or concepts have dominated German geography, most recently the *Neue Kulturgeographie* [New Cultural Geography] and the 'Cultural Turn' (Lossau 2008, Gebhardt, Reuber and Wolkersdorfer 2003). However, though many of the epistemological and methodological issues are identical or at least closely related, neither postmodernism nor theorizing 'the urban' has received much attention in these debates. This is both striking and strange given the levels of urbanization and the predominance of urban lifestyles in Germany and other (not only) western societies.

Yet, in spite of the widespread hesitation to accept the premises or theoretical insights of postmodernism, the empirical descriptions of urban development trends in Los Angeles and elsewhere have generally been accepted as insightful or at least interesting (e.g. Bell 2003, Ley 2001). Much in the same vein, Dear and Flusty's model of postmodern urbanism (Dear and Flusty 1998) has been interpreted and portrayed as a heuristic description of changing urban spatial structures even while largely neglecting the theoretical basis and context of the model (Kross 2006). Not least among these metatheoretical issues is the important question as to whether it is at all possible to analytically separate empirical description and interpretation or theorization of the urban.

It is not the purpose of this introduction to lament the lack of earnest theorization of what Michael Dear (from an American perspective) calls 'postmodern urbanism' within German geography. Rather, it is to recall these wide-sweeping descriptions and the connected theorizations as key reference points from which to explore some more narrowly focused issues related to the central themes of this book, which considers the restructuring of cities by non-'rational' processes of aestheticization and emphatic atmospheres. All the more so, as these theorizations stress the built environment of the city as highly relevant to our understanding and interpretation of social life and socialized atmospheres. Accordingly, the built environment occupies a prominent position in many current attempts to remake and reimag(in)e our cities – especially through highly visible, attention-grabbing developments that seek to fascinate and fuel the imagination of urban dwellers and actors. From such a vantage point, postmodern theorizations of the urban provide important insights and perspectives for understanding two interrelated aspects: firstly, the physical production of urban environments and how these are being perceived, and, secondly, how meanings are 'grafted onto' them, how we become fascinated or disgusted by them.

However, beyond focusing on the exceptional, mainly inner-city developments of internationally renowned 'starchitects', postmodern theories would have us pay special consideration to the (sub)urban periphery as well (Basten 2009). Understanding the (sub)urban periphery as a fringe or marginal zone, this is where the mundane and ordinary of everyday urban life worlds come into being, where meanings seem less clear, more fluid and also highly contested. Within this postmodern periphery,[1] aesthetics and the processes of designing urban places and spaces emerge as important and telling aspects of contemporary urbanity, at least as much as in 'globalized' inner-city locations.

Thus the first, more theoretical, part of this chapter will argue why it is important and relevant to look, firstly, at the (sub)urban periphery as a specific place, and secondly, at the design and aesthetics of the built environment – both in the city in general and in the periphery in particular. The second part of this chapter will then discuss the issue of perception. This is primarily based on empirical observations from my own research projects in Germany and Canada (Basten 2005). When the title speaks of perceptions *in* the postmodern periphery, the little word 'in' is supposed to express a certain ambiguity. It can and should mean both perceptions *from* the periphery as well as perceptions *of* the periphery. Images and imaginations of the periphery are highly diverse, not least depending on whether they have been created by 'insiders' or 'outsiders', by people who consider themselves at home there or by people who tend to regard the (sub)urban periphery as a strange and often alien world. In more theoretical terms, this sensitivity to perspective highlights the issue of plurality which, in a methodological as well as a normative sense, is very central to conceptions of postmodernity and postmodernism (Welsch 1988: 4ff.). Consequently, this second part of the chapter seeks to demonstrate what these different perceptions can tell us about space and place in the periphery today.

Postmodernism, Urbanity and Urban Life

Postmodernity and the Postmodern City

Michael Dear has differentiated three distinct meanings of the term 'postmodernism' or 'postmodern': as epoch, as method and as style (Dear 2000: 32ff.) – parallel to the ideas of context, method and concrete experience.

---

1 I use the terms postmodern periphery, (sub)urban periphery or simply periphery rather loosely and interchangeably to refer to what can also be described as the contemporary outer city. The theoretical background and connotations of these terms and their usage are discussed further below.

Starting with the latter, the dimension of style essentially describes certain attributes or conditions and their concrete experience. As such, it comprises 'distinctive cultural and stylistic practices' (Dear 2000: 5), in other words, particular patterns of behaviour as well as social and economic processes. It includes, but is not confined to the material consequences of such practices, as in 'an architectural style' or a particular aesthetic. Since nouns ending in the syllable '–ity' normally denote a condition or an attribute, it would seem helpful to use the term 'postmodernity' in exactly this way, that is to describe a set of conditions.

These very attributes or conditions are the elements that make us define 'postmodernism as [a separate] epoch' (Dear 2000: 34), a situational context with a certain *Zeitgeist* – the first of Dear's dimensions. In other words, the epoch is understood as a period of time during which postmodern attributes and conditions become notable, widespread or even dominant in their totality. Interestingly, in most academic contributions written in English, the noun postmodernity is now generally taken not primarily to denote the conditions but the epoch characterized by the predominance of these conditions, David Harvey's *Condition of Postmodernity* (1989) being the most obvious example. One could, of course, read 'the postmodern' (i.e., the noun) as the word that best describes this epoch dimension of the term. Arguably though, the use of the noun postmodern*ity* is not coincidental. Rather, it indicates that the material conditions and practices of postmodernity – style(s) if you wish – can to a certain degree, though not completely, be integrated into the epoch dimension of the term.

Dear's middle dimension between epoch and style, however, – postmodernism as method – seems rather different from the other two dimensions of style and epoch. This dimension is essentially what I prefer to reserve the term postmodern*ism* for. An '–ism' usually encompasses attitudes, world views, ideologies and the like – as well as epistemologies and methodologies – in other words, normative aspects rather than material ones or ones of practice. For that same reason, postmodern urbanity would seem a more appropriate term for most of what Dear calls postmodern urbanism, in particular, since such a distinction of terms becomes beneficial when analysing normative aspects like urban politics in the postmodern epoch.

Obviously, this categorical distinction between postmodernism and postmodernity remains problematic, since the two dimensions of method and epoch strongly interact. In practical terms the preponderance of postmodernist attitudes and practices that follow from them can be understood as characteristic of the postmodern epoch. And from a theoretical perspective the analytical categorization as 'either/or' is difficult anyhow, when seen from a postmodernist 'both/and also' ontology as favoured by Soja (Soja 1997a: 22).

Nevertheless, there are two good reasons for applying this distinction between epoch (and its styles) on the one hand and method on the other. The first is theoretical. If we consider postmodern*ism*, i.e. 'method' – or a particular version of it – as an ideology or philosophy, we can see it as a more or less logical system, internally coherent and consistent, but separate from material worlds or empirical observations of them. As such it is timeless – even though its emergence obviously is or has been contingent on material conditions – and thus it is very different from the postmodern as epoch. The postmodern as epoch is not, and cannot be, coherent or consistent, since it describes an empirical set of practices and conditions shaped by a heterogeneous multitude of forces and actors. What is central to the postmodern epoch, and postmodern urbanity or the postmodern city for that matter, is this very inconsistency, plurality, the simultaneousness and coexistence of postmodernist and modernist or traditionalist attitudes, of separate and contradictory styles, processes of production or reproduction. The postmodern city, then, cannot appear as a consistent whole, and it is more than just the particular cultural expression of a renewed late capitalism (cf. e.g. Jameson 1991, Harvey 1989), that is, it is not simply the outcome and result of a transformation of the economic base of society.

The second reason for using this two-fold distinction is pragmatic. It seems to be empirically useful since it allows for a multi-faceted approach to the empirical investigation of material conditions and processes in the city of the postmodern epoch, and thus it allows for debate about empirical findings across theoretical boundaries and divisions. Furthermore, it stresses the interconnections and interplay between attitudes, ideologies and methodologies on the one hand and material conditions and processes on the other as a key focus of enquiry, both in theoretical and empirical work.

The Postmodern Periphery

If we investigate the particular conditions that (re-)configure our cities in the postmodern epoch, a strong case can be made for shifting our focus of attention to the periphery, rather than maintaining the notable inner-city bias of traditional urban studies (cf. Dear's chapter in this book). It is a banal observation that the periphery has long become the normal and everyday living environment for an increasing majority of urban dwellers in most of our western cities. Furthermore, though the precise extent is highly variable, the old city centres have become increasingly irrelevant to those living in the periphery. These outer cities have been variously described as 'rather amorphous agglomerations' (Soja 1989: 188), as 'multinucleated metropolitan regions' (Gottdiener and Kephart 1991), or as a 'diffuse, unordered structure of completely diverse urban fields' (Sieverts 1997: 15).[2] These descriptions point to the observation that the periphery has undergone important spatial reconfigurations during the postmodern epoch, just as economic processes, social and cultural practices and values of the populations residing within it have also undergone significant changes.

Traditional notions and conceptualizations of the spatial (and economic and social) structures of our cities are strongly informed by the Chicago School of Urban Ecology and the models developed from its theoretical premises. Essential

---

2 'diffuse, ungeordnete Struktur ganz unterschiedlicher Stadtfelder.'

to these is the idea of a dominant core, which is not just a geometric focal point, but also the growth core of an urban agglomeration, providing the impulses and dynamics for the latter's expansion and differentiation. The core is thus the focal point of a strongly hierarchically organized space characterized by an uneven distribution of people, status, economic potential and, ultimately, power. This understanding of a centre, however, is inherently dualistic: it implies the existence of its opposite, a 'non-centre', an outer city relative to an inner city, a periphery relative to a core. In this way, theoretical conceptualizations of the modern industrial city of the late nineteenth and the first half of the twentieth century are also infused with a modernist logic of thinking based on dualisms and opposites, and the term periphery – developed from such a theoretical perspective – would be read as the opposite of (city) centre.

Such a dialectic understanding of periphery and core (or centre), however, no longer adequately grasps the realities of our contemporary urban agglomerations. Central business districts seem to be dissolving while new nodal points like edge cities are emerging in the outer city, leading to a general dissolution of the former opposites and the formation of something altogether new and different which seems to defy the traditional notions of city, centre and periphery (cf. Dear in this book). Conceptualizing the periphery in postmodern cities therefore becomes a potentially flawed undertaking. If we posit, as Dear has done, that in postmodern urbanism the periphery organizes the centre or that 'the direction of causality is from periphery to center' (Dear 2000: IX), we still maintain the very categories that a theoretical postmodernist critique of dualisms would rather dissolve. In other words, as long as the term periphery presupposes a centre to which it relates, and we dissolve the traditional notion of centre, we end up with a problem of definition (Hoffmann-Axthelm 1998: 112). To put it differently, if *Peripherie ist überall* [*Periphery is Everywhere*] (Prigge 1998) and everywhere is periphery, then there is no longer any real periphery.

A postmodernist conceptualization of the city thus needs to overcome the old dualisms of centre and periphery, of city and countryside, of city centre and suburbia, and should be open for the spatial and temporal inconsistencies and contradictions of process and pattern in the city. Hence the attempt to find new descriptive terms ('to come to terms') for these emerging postmodern urban spatial orders and structures. Edward Soja thus sees the postmodern city as 'exopolis', a parallel occurrence of the 'city turned inside-out' and 'the city turned outside-in' (Soja 2000: 250). From a similar, postmodernist perspective I use and understand the term periphery – retained partly for want of a better or more consistent word – or postmodern or (sub)urban periphery. A postmodernist understanding of periphery does not primarily read the term as the opposite of centre or core but rather stresses its meaning as edge, fringe or margin; not as the categorically different beyond a distinct border line, but as a space or area of interface and transition (Ellin 2000: 104). Sharon Zukin's idea of liminal spaces (Zukin 1992) or Tom Sieverts *Zwischenstadt* [*The In-Between City*] (1997) seem to stress the same idea. This kind of periphery, because it is arguably less inhibited by tradition and the built city of steel and concrete in the centres of modernity, is – or at least can potentially be – characterized by experimentation, diversity and fluidity in time and space, by a shifting and negotiating of meanings, by inconsistencies and contradictions. Thus, it could become the archetypical space of postmodern urban*ity*. It could also become the place to develop a new postmodern urban*ism*, an ideology appropriate to the postmodern city, since it involves not only the restructuring of space, but also the reconfiguration and redistribution of power, not least the power to define terms like centre/centrality or even city and civility.

Postmodern Urbanity – Aesthetics and Design

If we understand the postmodern epoch and the postmodern periphery – or the postmodern city as a whole – as a time and space of contradictions, plurality and fluidity, it becomes obvious that within it we will find a variety of different aesthetics, at least if we read the term aesthetics in the sense of a particular taste or style.[3] In terms of architecture and urban design, then, postmodernist designs will be found next to and side-by-side with modernist or traditionalist ones. Yet, very obviously, the importance of aesthetics and design in the postmodern city goes beyond the introduction of a postmodernist style and the resulting plurality of styles in architecture and urban design.

Firstly, the constructivist basis of postmodernist theory (re-)questions the possibilty of distinguishing between an external 'reality' on the one hand and perceptions, conceptions and images or representations of such a reality on the other (Soja 1997b). Since much of our orientation in space and our experience or interpretation of space is based on the sense of sight, any kind of image, picture or visual representation of the city forms a basis for communication and action. Thus such images and pictures become primary material for analysis and interpretation – to be understood in their respective contextuality and contingency.

Secondly, one of the observable trends of cultural change in the postmodern epoch is a widespread aestheticization of life. While aestheticization is not new to the postmodern city, its pervasiveness in many spheres of life arguably is. This is strongly linked to the economic rationales of capitalism and the commodification of culture and everyday life (cf. Jameson 1991, Harvey 1989). Yet, it is also linked to social differentiation and processes of identity-formation. Therefore, it is also notable in those spheres of life that are not, or are less directly, connected to the economic sphere. 'Theming' and 'branding' are just two particular outcomes in this respect. Both can be seen as strategies for establishing and for using specific aesthetics for particular purposes (cf. Sorkin 1992). This becomes notable in urban

---

3   Clearly, the concept of aesthetics, especially if a wider understanding of the term is used, embraces the distinction between attitude, philosophy or ideology on the one hand and material style on the other. It thus illustrates the interplay and interconnectedness of the different dimensions of the postmodern.

development, urban politics and urban life in general – not just in a commercial market-place.

One specific result of this situation is the more conscious and deliberate use of architecture and urban design in all aspects of urban development, since such design strongly influences the perception and conception of images and representations of the city. This is very notable with flagship projects of urban renaissance or renewal in inner-city locations, which have received much attention in this respect (Heeg 2008). However, it is also particularly notable – and not just coincidentally, I would argue – in the postmodern periphery. Thus I would like to explore this aspect a little further, turning to some empirical investigations of perceptions in the postmodern periphery.

Perceptions in the Postmodern Periphery

Let us think about suburbia for a moment. Let us reflect upon our immediate associations, our visual perceptions and the links between those images and our notions of life 'out there'.

What does 'our' suburbia look like? What kind of design solutions do we see before our eyes? Are we possibly seeing the uniformity of Levittown[4] or of modern tract developments of identical houses? Or are we seeing detached single family houses in Germany which, though individually built, appear paradoxically uniform in design – each one a family's dream home to be handed down through generations rather than seen as an investment or a commodity to be traded in every five years? Or are we seeing suburbia as an area of innovative design, of uniqueness, individuality and difference, a space in which to do what one wishes in one's own home, and thus as a symbol of individual freedom?

Yet, to what degree are these perceptions of ours based on personal experience of life in suburbia? Could it be that these images are primarily based on media images and representations of suburbia on the one hand and on elitist and academic discourses on the other? Perceptions from within the postmodern periphery – of those who dwell there – can be decidedly different from those of people who perceive it from the outside, some of whom may be involved in the construction or design of images of and for this periphery (cf. Warnke and Gerhard in this book).

---

4   Levittown, near New York, is the archetype of an early post-war 'master-planned' suburban community relying on the standarization of (very few) house designs and mass-produced, prefabricated materials. This allowed for very cheap housing, but created a physical, social, ethnic and racial uniformity over vast tracts of suburban land. The company, Levitt & Sons, went on to build other Levittowns in Pennsylvania, New Jersey and even Puerto Rico.

Perceptions from Without

It is easily possible to argue that elite discourses, particularly among architects and urban designers, but also among planners and academics, for a long time largely ignored suburbia, seeing it as a non-place, especially in aesthetic terms. By and large, suburbia held no fascination for planners, architects or urban designers. The exact reasons for this are certainly different from country to country, but the observation, I think, holds true for North America as well as for Germany and other countries in Europe. However, this situation has changed over the last two decades. Design issues, both of architecture and of urban design, have recently received far more attention in new development projects in the periphery. Which is not to say that the suburban periphery was not a designed space or place before – clearly, early suburbia/exurbia was (cf. Fishman 1987) and post-1945 suburbia was as well – but post-1945 discourse about urban design largely tended to ignore the (sub)urban periphery as simply unaesthetic. Slowly but surely, though, the periphery has started to be perceived not just as a land use planning problem, an ecological problem or a socio-economic (segregation) problem. It has also started to be seen as a design challenge. The reasons for such a change in perspectives tend to be different for the different groups of actors involved.

Figure 5.1   Suburbia as design challenge. A street scene in the postmodern suburban periphery: Rather than lamenting its ugliness – or the socio-economic exclusiveness of the project as well as its architectural expression – it could also be viewed as a design challenge.

Particularly in the North American context, the developers play a pivotal and politically powerful role in suburban development. They study and test markets, consumer tastes and attitudes, they observe aesthetic trends and styles and develop their products accordingly, and they try to influence demand by theming and branding strategies or advertising campaigns (Roost 2000). In this sense, they are no different from producers of other consumer goods. If they have increasingly turned their attention to design issues, it is because they believe that design sells – good design, design with a difference etc. – and that the importance of design for home sales is greater today than in the past. What currently sells very well, not just in North America, is what can be called 'retro design'. This refers to a somewhat eclectic mix of traditionalist architectural or design styles, alluding to quality craftsmanship and social or family values. It mostly combines superficial design elements with the latest modern interior design and technology (Duany 2000). The only element missing for it to be described as truly postmodern architecture is, I suppose, the sense of irony and the self-reflexive playfulness.

Many such 'retro design' projects make use of image and marketing strategies which deliberately play with emotions evoking a sense of security, homeliness and community or, as their negative opposites, of fear (of others, of crime etc.), anonymity or isolation.

Highly relevant for the comprehension of the postmodern periphery is the fact that the production of the built environment – something quintessentially material – more than ever now starts with the creation of images that do not *re*present material worlds. Instead, these images support and direct the creation of imagined worlds, often imagined past worlds, establishing a market for material products that are developed through fostering emotional and normative connotations. Here, values such as community, neighbourliness, family and sanctuary – and their opposites: fear, anonymity and isolation – become attached to images which are materialized in both architecture and urban design. Therefore, the housing industry, especially in North America, is becoming more similar to the film industry, and the Disney company's 1990s move into urban development in the town of Celebration, Florida, is emblematic in this respect. As reality and virtuality seem to become indistinguishable and inseparable, there's simply no analytical possibility left to tell what life in suburbia or small town America was or is really like (Roost 2000, Ross 1999).

These are not only concerns of the developers. Planners, local politicians, architects and urban designers are also involved in the creation of the built environment and thus participate in the creation of images and perceptions for the postmodern periphery. They also have to pay heed to the economic imperatives which dominate the developers' view of suburban development, but not exclusively so, and not at all times and in all places. Local power structures, nimbyism, smart growth initiatives, civic boosterism, individual values and beliefs all influence political agendas and may generate a stronger consideration of design issues. Planners, just like society at large or at least some prominent lifestyle groups, have often adopted certain 'post-materialist' values and attitudes that correspond to new ecological or aesthetic imperatives. These in turn contribute to the architectural and urban design challenge which the periphery poses today. Certainly, if the periphery is where things are happening, and if the importance of conscious design decisions in the production of the urban built environment has greatly increased, then the periphery deserves our special attention in theorizing contemporary urbanity and the powers that give shape to it.

Power is relative and contextual. The powers of the market are never absolute, they may be constrained and countered by the power of the state – in its various guises – as well as by the power of other influential groups in society. The contingency of specific power constellations has been clearly revealed in my own analyses of new peripheral development projects in Germany and Canada in the 1990s (Basten 2005). In general, it can be said that those actors who are not developers tend to be more important in projects in Germany than in North America. Yet, the local and project-specific constellation of decision-making powers is highly variable.

For example, it appears that architects and urban designers have conceptions of their role and importance within the 'production process' of the built environment that are both highly individualistic and socio-culturally contingent. These conceptions are shaped by processes of professional training, institutional socialization, and socio-cultural discourses on the relationship between architecture and society. This poses several questions: Does an architect see him- or herself as an artist who creates an aesthetically and socially ideal world in miniature? How does an architect or urban designer position him- or herself between the state bureaucracy on the one hand, and the developer on the other? Do architects (or planners, or politicians) regard the postmodern periphery as a central challenge for tomorrow's urban places? These are only a few of the questions that may have an influence on the project-specific constellation of power relations between the different actors involved in shaping the (sub)urban periphery.

Perceptions from Within

Not surprisingly, those who live in the periphery tend to develop perceptions of the periphery that are rather different. Theirs is first and foremost the everyday life perspective of a dweller, in other words, it stems from lived experience, which is not always reflected rationally, but includes affective dimensions and evaluations (like fascination or boredom). Such perceptions of the periphery turn out to be quite distinct from the more abstract, often ideological perspectives of professionals, especially design professionals. This is not to say that design is not important for the residents of the periphery, but that the *aesthetic* dimensions of design tend to be less relevant to them than the pragmatic dimensions. So, design should be based on, or at least be mindful of, the dwellers' perceptions rather than the developers'.

Design, in this context, is important, firstly because it allows a better, easier and more pleasant arrangement of life, in other words, because of its functionality

– or disfunctionality, as the case may be. Specific design solutions, e.g. scale, design elements like porches or balconies, the accomodation of car-traffic etc., are generally experienced as making everyday life more or less practical and pleasant. Rarely are they experienced as ideologically loaded or perceived as inherently positive or negative for social practice. Thus, everyday life is mostly not a self-reflexive political or ideological sphere.

Secondly and more significantly, however, design is important for creating distinguishable places. Attention to design in the periphery can facilitate better readability of that 'diffuse, unordered structure of completely diverse urban fields' (Sieverts 1997: 15)[5] that is the postmodern periphery. Frederick Jameson has described a postmodern hyperspace transcending 'the capacities of the human body to locate itself, to organize its individual surroundings perceptually, and cognitively to map its position in a mappable external world' (Jameson 1991: 44). And David Harvey has interpreted the adoption of neotraditionalist architecture in postmodern suburbia as an escapist counter-reaction to an existential loss of security (Harvey 1997). It would seem, however, that most residents of the postmodern periphery clearly conceive the periphery at large as a multi-centred space in which they do not expect to find places that conform to traditional notions of rootedness and locality – even if medially reproduced or created images of such places are common knowledge among them. Though they know the periphery to be unlike traditional city centres or small towns, they still feel themselves equipped to emotionally create their home within it. To put it differently, many users of Jameson's postmodern hyperspace already seem to have acquired significant capabilities of cognitive mapping and reading these new texts of postmodern urban landscapes without necessarily developing schizoid or pathological tendencies. In this sense, psychological rootedness in the periphery requires distinguishability and readability of its landscapes, and this in turn calls for more attention to design, though not for specific prefigured design solutions and architectural styles.

Closely related to this aspect is the importance of design to act as a means of distinction. Design helps residents to create in their minds not only a topographical map of the periphery, but also a social map representing socio-economic and cultural boundaries within it. In this way, design helps to distinguish between 'us' and 'them', reflecting the 'de facto' heterogeneity of the periphery. Different styles of architecture and urban design (retro or modern) are one particular means to achieve such distinctions, and perceptions of physical and material differences are being connected to perceived socio-economic differences. So far, there is a clearly observable search for relative socio-economic homogeneity within one's more immediate home environment as such homogeneity seems to support the ability to feel at home and to experience home as a place of sanctuary. More existentially, it supports identity formation and sense of self, just as it has done throughout modernity. This contradictory tendency makes the periphery as a whole a differentiated, heterogeneous patchwork in space, with the result that the

---

5   Cf. footnote 2, above.

perception and construction of a socio-economically differentiated space is often reproduced like an ecological fallacy: people from elsewhere are seen as different, because they live in another part of the periphery which looks different.

The perception and construction of the periphery as a differentiated, patchwork-like social space – by its residents as well as its 'producers' – thus closely reflects general observations of an increasingly differentiated and plural society at large. Within this patchwork of the postmodern periphery, the place and time of one's identity tend to become smaller: the 'home' patch becomes more localized and confined even while the dwellers' activity patterns become more extensive and fragmented in space. Simultaneously, it becomes more narrowly bounded in time, as moving from one stage in life to the next almost invariably implies moving house and neighbourhood, particularly so in North America. Therefore, the postmodern experience of fragmentation seems to be more common and characteristic for the periphery than for the urban centre, forming the basis of a different type of contemporary urbanity.

Figure 5.2   **Distinction through theming. Conscious use of (in this case retro) design and theming in design and marketing allows for the creation of a distinct project, a recognizable address 'out there' in the postmodern suburban periphery. This can aid the re-formation of individual and group identities – of course, at the cost of excluding others.**

Source: Basten 2005: 217

A Postmodernist Conclusion

Both from a theoretical and an empirical viewpoint, then, the periphery warrants our close attention. It is here that most of the normal and everyday construction of identity and society takes place – in a twofold sense of 'taking place': occuring as well as occupying and shaping. Consequently, the periphery is the space where the most intricate landscapes of the postmodern city are being shaped through consciously designed, but often unconsciously experienced, materialities, images and representations.

It is worth remembering, though, that everyday life perspectives are never unideological or unpolitical per se. So if we emphasize all the different ways in which residents of the periphery perceive and construct their social worlds and inscribe their meanings on the periphery, this does not mean that we wish to deny the importance of structural factors and power relations. The often seemingly anonymous activities of real estate markets and the political systems that regulate development are just two of the most obvious examples of such structural factors which significantly influence the shaping of the periphery and the life worlds within it. Likewise, the abstract and ideologically loaded professional or academic discourses are not at all irrelevant, whether they are about the importance and meaning of the periphery vis-à-vis the 'old' centres or the merits of specific design solutions or styles.

Of course there is an economic as well as a political side to this, and there is a highly uneven distribution of power in the resulting processes of decision-making. Yes, the increasing importance of aesthetics and design in the periphery is a symptom of the commodification of design in everyday life (cf. Jameson 1991, Harvey 1989). It is another facet of the late-capitalist restructuring of the housing market to secure profit, while partially hiding the profit motive behind a discourse on aesthetics as well as on social and family values. Bread and games. It is that, but it is also more than that – or it can be more.

Conceptualizing the periphery as a postmodern space in the making is a theoretical (and a political) challenge just as designing (for) the periphery is an aesthetic challenge that also refers to the lifeworld dimension. I have always been intrigued by the strong parallels between certain critiques of postmodernism as theory and the predominant critiques of one particular attempt to re-design contemporary suburban landscapes, i.e. a retro-design-oriented New Urbanism. Repeatedly the same charge is levelled against practically everyone who truly tries to tackle the challenge of designing (for) the postmodern periphery: escapist, fassade-oriented, superficial, possibly demagogical, in any case reactionary (e.g. Marcuse 2000, Soja 2000: 248, Harvey 1997). Can there though be anything more reactionary than maintaining a status quo by ignoring the challenges at hand? Imagining a revitalized urban aesthetic is not enough if this simply means revitalizing the aesthetic of the modernist centre and trying to copy it in the periphery. But attempts at designing truly postmodern*ist* places that allow for uncertainty, change, heterogeneity, ageing in situ etc. seem to be few and far between. Maybe it would be more progressive in this context to begin to perceive the periphery not just as an aesthetically flattened pastiche, but as a realm of possibility, opportunity, fluidity and hybridity – in the making and in our making.

References

Basten, L. 2005. *Postmoderner Urbanismus. Gestaltung in der städtischen Peripherie.* Stadtzukünfte, vol. 1, Münster: LIT Verlag.

————— 2009. 'Überlegungen zur Ästhetik städtischer Alltagsarchitektur'. *Geographische Rundschau*, 61(7–8), 4–9.

Becker, J. 1997. *Geographie und Postmoderne. Eine Kritik postmodernen Methodologisierens in der Geographie.* Beiträge zur Kritischen Geographie, vol. 1. Wien: Österreichische Gesellschaft für Kritische Geographie.

Bell, T.L. 2003. 'Review of: *Postmetropolis: Critical studies of cities and regions* by E.W. Soja'. *Annals of the Association of American Geographers*, 93(1), 248–50.

Dear, M. 2000. *The Postmodern Urban Condition*, Oxford, Malden: Blackwell Publishers.

Dear, M, and Flusty, S. 1998. 'Postmodern Urbanism'. *Annals of the Association of American Geographers*, 88(1), 50–72.

Duany, A. 2000. 'Eine Gegenrede: Nichts als Vorurteile!' *Stadtbauwelt*, 91(145), 36–41.

Ellin, N. 2000. 'The Postmodern Built Environment', in *Design Professionals and the Built Environment: An Introduction*, edited by P. Knox and P. Ozolins. Chichester: John Wiley & Sons, 99–106.

Fishman, R. 1987. *Bourgeois Utopias. The Rise and Fall of Suburbia*, New York: Basic Books.

Gebhardt, H., Reuber, P. and Wolkersdorfer, G. (eds) 2003. *Kulturgeographie. Aktuelle Ansätze und Entwicklungen.* Heidelberg: Spektrum Akademischer Verlag.

Gottdiener, M. 1994. *The New Urban Sociology*, New York: McGraw Hill.

Gottdiener, M. and Kephart, G. 1991. 'The Multinucleated Metropolitan Region: A Comparative Analysis', in *Postsuburban California: The Transformation of Orange County since World War II*, edited by R. Kling, S. Olin and M. Poster. Berkeley, Los Angeles, Oxford: University of California Press, 31–54.

Harvey, D. 1989. *The Condition of Postmodernity. An Enquiry into the Origins of Cultural Change.* Oxford, Cambridge: Blackwell Publishers.

————— 1997. 'The New Urbanism and the Communitarian Trap'. *Harvard Design Magazine*, 1(1), 1–3.

Heeg, S. 2008. *Von Stadtplanung und Immobilienwirtschaft. Die 'South Boston Waterfront' als Beispiel für eine neue Strategie städtischer Baupolitik.* Bielefeld: Transcript.

Himmelfarb, G. 1997. 'Telling It as You Like It: Postmodernist History and the Flight from Fact', in *The Postmodern History Reader*, edited by K. Jenkins. London: Routledge, 158–74.

Hoffmann-Axthelm, D. 1998. 'Peripherien', in *Peripherie ist überall*. Edition Bauhaus, vol. 1, edited by W. Prigge. Frankfurt am Main, New York: Campus, 112–19.

Jameson, F. 1991. *Postmodernism, or, the Cultural Logic of Late Capitalism*. London, New York: Verso.

Kross, E. 2006. 'Modelle im Geographieunterricht: das Beispiel der lateinamerikanischen Stadt', in *Kulturgeographie der Stadt*. Kieler Geographische Schriften, vol. 111, edited by P. Gans, A. Priebs and R. Wehrhahn. Kiel: Geographisches Institut der Universität Kiel, 491–508.

Ley, D. 2001. 'Review of: *The Postmodern Urban Condition* by M.J. Dear.' *Annals of the Association of American Geographers*, 91(3), 577–79.

Lossau, J. 2008. 'Kulturgeographie als Perspektive. Zur Debatte um den cultural turn in der Humangeographie – eine Zwischenbilanz'. *Berichte zur Deutschen Landeskunde*, 82(4), 317–34.

Marcuse, P. 2000. 'The New Urbanism: the Dangers so far'. *DISP*, 36(140), 4–6.

Prigge, W. (ed.) 1998. *Peripherie ist überall*. Edition Bauhaus, vol. 1. Frankfurt am Main, New York: Campus.

Roost, F. 2000. *Die Disneyfizierung der Städte. Großprojekte der Entertainmentindustrie am Beispiel des New Yorker Times Square und der Siedlung Celebration in Florida*. Stadt, Raum, Gesellschaft, vol. 13. Opladen: Leske + Budrich.

Ross, A. 1999. *The Celebration Chronicles: Life, Liberty, and the Pursuit of Property Value in Disney's New Town*. New York: Ballantine.

Sieverts, T. 1997. *Zwischenstadt. Zwischen Ort und Welt, Raum und Zeit, Stadt und Land*. Bauwelt Fundamente, vol. 118. Braunschweig, Wiesbaden: Vieweg.

Soja, E.W. 1989. *Postmodern Geographies: The Reassertion of Space in Critical Social Theory*. London, New York: Verso.

———— 1997a. 'Six Discourses on the Metropolis', in *Imagining Cities: Scripts, Signs, Memory*, edited by S. Westwood and J. Williams. London, New York: Routledge, 19–30.

———— 1997b. 'Planning in/for Postmodernity', in *Space and Social Theory: Interpreting Modernity and Postmodernity*. The Royal Geographical Society with the Institute of British Geographers Special Publications Series, vol. 33, edited by G. Benko and U. Strohmayer. Oxford, Malden: Blackwell Publishers, 236–49.

———— 2000. *Postmetropolis. Critical Studies of Cities and Regions*. Oxford, Malden: Blackwell Publishers.

Sorkin, M. 1992. 'See You in Disneyland', in *Variations on a Theme Park. The New American City and the End of Public Space*, edited by M. Sorkin. New York: Hill and Wang, 205–32.

Welsch, W. 1988. *Unsere postmoderne Moderne*. Weinheim: VCH Acta Humaniora.

Zukin, S. 1992. 'Postmodern urban landscapes: mapping culture and power', in *Modernity and Identity*, edited by S. Lash and J. Friedman. Oxford, Cambridge: Blackwell Publishers, 221–47.

PART II
Focusing Fascination:
Crossing Theoretical and Empirical
Perpectives

Chapter 6

'The Most Dangerous Knack': Fetish and Fascination in the Built Environment

Neil Smith

… because we are prone to accept appearances for reality, … for each art there is some knack that imitates its effects and misleads its victims … . Gymnastics correlates with cosmetics, which gives the appearance of health without the reality. Rhetoric, the power to substitute appearance for reality in language, is the supreme and most dangerous knack.

(Feenberg 2001: 135).

Introduction: Appearance and Reality

All societies build their own deepest social interests, the deepest social secrets about their inner workings, and the sources of their greatest fascination into the landscapes they craft. In the repetitive normality of daily life these truths are either hidden in plain view or else vaunted as spectacle. Medieval castles built on high ground above walled cities, surveying the surrounding countryside, bespeak the potency of military power in that era; Christian church spires, Muslim minarets and the Buddhist temples of Lhasa evoke (or evoked) the religious power constitutive of other equally specific social and historical geographies; in like fashion the skyscraper forests of contemporary capitalist cities, no longer just in the downtowns, are forbidding monuments to corporate class power. Such seemingly prosaic symbolic landscapes are every bit as symptomatic of the societies that built them as are more monumental spectacles: to name but a few, those to national power (for example the National Mall in Washington D.C. or the Forbidden City in Beijing); more flamboyant paeans to consumptive capitalism such as the various Disney theme parks; or the amalgam of nature and nation represented in 'national' parks.

Written into the visible landscape, these social secrets, interests and fascinations are naturalized – fetishized – refracted back as the most unassailable facts of social existence. The appearance of the built environment thereby veils, even as it expresses, its social realities. If fetishism for Marx meant that the relations between people come to appear as the relations between things (commodities), then spatial fetishism confuses the appearance of places for the relations between the people who constitute different places: the social relations between people come to be encompassed as the relations between places (Anderson 1973). This spatial fetishism lies at the heart of today's geographical ideologies, whether expressed in the romancing of specific places, the metaphorical pairing of apparent opposites such as city and countryside or city and suburb. It also molds the synecdochal conflation of the people and social relations that comprise nations – 'Iran', say, or 'the United States' – with the governmental apparatuses embodied in states which dominate these places; that is, it confuses the relations between and among people with the relationship between institutions and territory. Insofar as the reflexive power of these very real people-built places comes to demand or attract social identification with the place itself, as for example with nationalism, national identities, or other forms of localism, here indeed is an ideological rhetoric that involves 'some knack that imitates its effects and misleads its victims' (Feenberg 2001: 135).

This language of appearance and reality is not much in fashion today in an intellectual era still sobering up from a collective dalliance with postmodernism, from the consequent subsumption of anything approaching the real under a fascination with the sign, and from the quest to deconstruct the sign. At the very least the disjuncture of real and apparent, which was so productively if quite differently engaged by a range of thinkers, from Kant and Hegel to Marx and Freud, is sublimated. The distillation of much intellectual work today around the discourse of post-structuralism, which is far more sustainable than postmodernism, does not indulge so self-evidently in 'fascination' and is even less obviously concerned with 'economy'. Indeed in English-speaking intellectual circles post-structuralism often performs as a certain anti-economics. And yet I want to suggest that these silences concerning economy and fascination are themselves symptomatic of a certain knack coiled within the deployment of 'discourse'.

With its often ambivalent commitment to the dialectic, much post-structuralist theory nonetheless posits talk of reality as a crude positivist fantasy: only through discursive practices can reality be accessed, the argument goes, and those discursive practices therefore become the simultaneous nexus and substance of social and cultural theory. Reality here is at best out of focus; at worst the very real problems of access disqualify non-discursive reality as a plausible subject for serious social theory. Let me make this concrete with an example. Attempting to tackle the question of human involvement in the making of nature, geographers Noel Castree and Bruce Braun (1998: 7) derive from the insights of post-structuralism a 'deep discursivity' according to which 'nature is constructed "all the way down"'. Whatever the accompanying caveats, such an effective excision of reality owes ultimately to a radical Kantianism which is not only inadvisable but presumably unsustainable. The ultimate unknowability of the object-in-itself, in Kant's terms, leaves no alternative but an idealism of the object-for-itself. Here scientists among others, casting a glance toward cultural theory in the last two decades, have, whatever the contradictions inherent in established scientific practice itself, had little trouble in debunking such claims in the name of common sense.[1]

The language of appearance and reality owes most recently to Marx, whose theory of ideology pivoted on the distinction between the appearance of the commodity and the social relations it embodied. For Marx, the dialectic of appearance and reality represented a riddle not of philosophy but of historical, social practice. Social and political analysis had no alternative, Marx argued, but to interrogate the appearances of the world, rather than taking them for granted, as a means of eliciting the more encompassing real. Herbert Marcuse had something very similar in mind, as did Henri Lefebvre. 'Reality' Lefebvre defines as 'this singular mix of the concrete and the abstract, signs and things, truths and illusion, static appearances and dynamic lies' (Lefebvre 2009 [1986]: 271). The dialectic of appearance and reality naturally has an older philosophical source. For Hegel, appearance and reality are united in the final victory of the concept, much as the Subject fuses with the Object. And Marcuse's own musings were actually a reflection on Plato.

But it is to Kant's philosophy that contemporary post-structuralism returns with its suspicion of the real. Kant's eventually unsuccessful attempt to resolve the dialectic of appearance and reality instead instantiated a radical separation between the two, his theorization of the phenomenon notwithstanding. For German philosopher, Johann Gottlieb Fichte, this separation represented a certain *hiatus irrationalis* in Kant, an ultimately unbridgeable gap between concept and reality.[2] Or, in the language above, a social construction 'all the way down' which can never approach reality. One can recognize clear traces of this dilemma in Foucault, to take just one obvious example, whose substantially metaphorical 'micro-geopolitics' of social space implicitly mobilizes a Kantian inaccessibility of space per se as a priori and, following Newton, effectively absolute (Smith 2000).

The irony in all of this, of course, is that while reality is in this way written out of the social theoretical script, or at least held at arm's length, 'ontology' becomes a central conceptual fulcrum for this same post-structuralist theory. The inaccessibility – indeed the effective un-nameability – of reality makes it an even more prized acquisition. Insofar as different individuals, theories, purviews and approaches are seen to possess or express their own ontologies, reality is rehabilitated only at the expense of its relativization. This therefore implicitly reinvents a fascination with reality, quenched by reference to new ontologies. That fascination is perhaps most obviously manifested in the new empiricism of much English-language cultural geography today, refracted through the idealism embraced in the effective refusal of reality. As critiques of scientific method suggest, the mirror image of idealism is an empiricism emanating from the other side of the *hiatus irrationalis*, and so

---

1   See for example the Sokal affair, in which a scientist spoofed the social sciences and humanities by proposing an entirely nonsensical, invented, constructivist reading of quantum mechanics (Sokal 1996).

2   Cf. Smith 1989; for a broad re-evaluation of Kant on geography, see Elden 2009.

substantive research in this vein often fastens on specific empirical appearances – objects, events or processes – in the cultural landscape that are fertile for exploration and/or deconstruction in search of 'ontologies of difference'.

Fascination with cultural appearances here obscures the dialectic of appearance and reality or else re-establishes a quite undialectical relationship expressed in a more or less passive constructionism. That which was swept out through the front door, namely reality, is eagerly embraced through the back door. The dismissal of reality, with a capital R, is here compensated by the regrounding of post-structuralist epistemology and a certain 're-appearance' of reality in the language of ontology – ontology again as ideology.

This way of presenting the question of appearance and reality will inevitably draw criticism. It will be seen as an attempt to reinstate some privileged access to an unproblematic reality. Nothing could be further from the truth. I would contend that such a fear of a revived positivism makes any kind of logical sense only if one presumes from the start the Kantian *hiatus irrationalis* that irreparably divorces reality from the means we have of knowing it – reality from its appearance. But what if we were to take our starting point from a Hegelian unity and try to explain instead how a separation of appearance and reality could have claimed such conceptual power in the first place? Stepping back from the implicit abandonment of the real does not require any reinstatement of presumed privileged access to reality; even less does it presume any necessary simplicity inherent within that reality. Access to the real, in fact, is always a matter of intense social struggle, and what we take as real is just as complicated a product of social (including but far from exclusively representational) struggles – struggles which are lost as soon as such representations are conceptually isolated from what they purport to represent. Reality is quarantined. Rather, I am more interested in asking how the question gets framed this way, and whose interests such a framing serves; I hope to expose and yet step back from the abandonment of reality implied in some strands of contemporary social theory.

Fascination and Ideology

In an attempt to make some of these abstract claims more concrete, let me examine two specific issues in our current understanding of the built environment. I am tempted to try to illustrate this argument through the gentrification literature and the evolution of urban gentrification from its rather marginal beginnings into a global strategy for city building. This would allow us to survey scholarly analyses of the gentrification spectacles that increasingly punctuate the urban landscape, focusing simultaneously on questions of economy and fascination. It would also provide for a more concrete demonstration of the claim that much post-1970s cultural geography increasingly sports a revived empiricism. But there is also the danger that such a discussion, while relatively straightforward, would inevitably rehearse some older arguments that are probably well known by now. Instead, therefore, I propose to focus briefly on two more searching issues that may not immediately connote questions of economy and fascination: first, the question of 'urbicide' and the new techno-military urban landscapes of the twenty-first century; and second, the recurrent but again vital question of the state in this post-structuralist historical moment.

The Techno-military Landscape

In a pathbreaking volume of essays and several subsequent articles, British geographer Stephen Graham (2004, 2007a, 2007b) has argued that warfare in the twentieth century is an increasingly urban affair. To an unprecedented extent, he suggests, cities are today the targets of state-sponsored and non-state warfare: war, terrorism and cities are mutually redefining each other in a process of creeping 'urbicide', by which he means 'the deliberate denial, or killing, of the city' (2004: 25). This new urban geography of the battlefield is both the product of and inspiration for a whole new generation of military technologies which take cities of the global south as their presumed target and testing ground. This panoply of novel military technologies is more deeply rooted in a whole new techno-industrial economy of security and surveillance generated in the US, Israel and Europe, as well as many other places, which were clearly in existence before but have burgeoned since 2001. In the battlefield, explains one US major, the point is to 'create operational shock in the urban environment' (quoted in Graham 2007b: 17).

In astonishing detail, Graham investigates the ways in which 'US military planners imagine and discursively construct global south cities as their predominant "battlespace" for the early twenty-first century.' He provides a tour through the new, imminent and proposed technologies emanating from multiple alphabet agencies of the US military but also a whole shadow world of networked corporations, research institutes and training centres that comprise a new 'military-industrial-communications-academic complex'. In one example, he records the work of the US 'Combat Zones That See' project (CTS) which, from pilotless drones in the sky, seeks to discriminate between the 'normal' and 'abnormal' individual movement of people in the urban environment. This project seeks to build up 'representative data profiles' which, via mathematical algorithms, can identify abnormal movements, targeting them instantaneously. In another venture, US military technologies would identify, 'within a fraction of a second', the source of any gunfire and respond automatically from the same drones, without any human mediation (Graham 2007b). Collateral damage is not an unfortunate side effect but an integral result, even an intended one, of such techno-military apparatus.

In a parallel work, British-Israeli scholar Eyal Weizman (2007) carefully unfolds the architecture and strategic geography of Israeli occupation in the Palestinian territories. Settlement geography *is* military geography in this case and Weizman is meticulous in unpacking the spatial strategies not just of the military but of the state per se. He documents the architectural strategies first laid out decades ago by then-prime minister Ariel Sharon, and the vertical as much as horizontal strategy of occupation this produced, and he examines the power of what he calls 'optical urbanism' in the settlements on Palestinian land, and the extraordinary surveillance technologies of military checkpoint design. He traces the impossible territorial violence of the Israeli wall, which is designed to isolate and control 'a permanently temporary Palestinian state', and the technologies of targeted assassinations. Most revealing, he records the strategic, tactical and organizational work of the Operational Theory Research Institute (OTRI) in Israel, headed by two retired generals, which pursued plans for 'low intensity' or 'dirty' wars. In that effort they explicitly deployed the work of Deleuze and Guattari, in one case using their metaphorics of space to develop the tactic by which Israeli soldiers 'walked through walls' (Weizman 2007: 185). That is, rather than move into Palestinian towns along the streets, where soldiers were highly visible and vulnerable to attack, they simply blasted through Palestinian homes, advancing into neighbourhoods parallel to the outside streets.

The work of Graham and Weizman and numerous others in constructing these techno-military landscapes provides a tremendously revealing – and certainly fascinating – entrée into the intensified militarization not just of war but of everyday life. These accounts are at their best in depicting the extraordinary flexibility, experimentalism and constant mutability of techno-military strategies. In their different ways, they stand as sharp, shocking indictments of the military-industrial-technology complex and its multifaceted productions of space in the name of state power. The mirror image of this spectacular militarization of everyday life, it has to be pointed out, is the explicit economic rationale entwined with the military. In Iraq after 2001, for example, the number of corporate mercenaries – euphemized in political and media ideologies as 'private contractors' working for corporations such as Blackwater (since renamed Xe Services), Bechtel and KBR (Kellogg, Brown and Root) – quickly outnumbered the military personnel, illustrating the fact that capital accumulation and the opportunistic pursuit of neoliberal reconstruction were an integral part of the rationale for the military intervention in Iraq (Klein 2007; Scahill 2007).

The work of Graham and Weizman provides vital evidential ammunition for resistance struggles against such intensified and corporatized militarism. But in their different ways, each of these accounts also becomes so absorbed in the immediate 'scandal' of such technologies and strategies that the question of *explaining* these shifts never quite comes into focus. This is perhaps more evident with Graham's research. For example, the claim that warfare is now about cities in a way it never was before, seems a historically dubious generalization; the sacking of cities, as the seats of religious, military and economic power, was a staple of pre-capitalist wars. Is it possible that this exaggeration is prompted by a techno-military fascination which the author (and, it should be said, we as the audience) shares with his objects of research? Moreover, is it cities that are the true targets or is it the people living in them? Graham freely concludes that many of the techno-military plans he examines may never actually come into effect. They may be 'wild fantastical discourses' (2007b: 33), he says, and indeed there are many in the military who are themselves sceptical. As Graham is well aware, this necessarily leaves us questioning the status of the discourse analyses that comprise the story: if these may or may not be idle fantasies of military techno-geeks, how seriously are they to be taken, how are we to decide, and how far should our fascination stretch in pursuit of analysis? Answering these questions presumably calls for a more thorough contextualization of these techno-military discourses and especially an understanding of the various social interests to which they respond and the struggles between different interests that shape ideological discourses.

Likewise, the Israeli OTRI was actually disbanded in May 2006, and while Weizman's analysis is more thoroughly contextualized within competing Israeli and Palestinian interests, it would be important to know the significance of the OTRI's disbandment and to have a more replete map of competing social interests within Israel and its military, and in the occupied territories. This is important because it begins to point a way beyond simply scandalizing the detached efficiency (or otherwise) of military technology in controlling entire populations and in killing people designated as 'targets'. The danger here is that fascination with the technology will not so much provide a means to oppose the 'shock doctrine' of capitalism, as Naomi Klein (2007) calls it, but will, rather, simply produce a mirror image, an oppositional shock doctrine that feeds fear of the state as much as it feeds any mobilization against it. If, as Klein argues, the creation of shock and disaster has opened up whole new terrains for capital accumulation (see also Smith 2007), how are we to ensure, as well as can be done, that shocking revelations by political progressives concerning the new techno-economics of killing do not themselves inadvertently fuel that which they would oppose?

The State

In popular parlance, and even increasingly in academic writing, the state is often equated simply with government. Even where this is not strictly the case, liberal theories of the state treat it as either an outright abstraction or else as a reified object, a collection of institutions identifiable as separate from civil society and the economy. Against such treatments of the state, and contrary as well to Marxist considerations of the state as the political and institutional pupa of class relations, much post-structuralist discussion has variously sought to de-centre the state as any kind of material entity wielding social power. Foucault, for example, was much more interested in locating power among the interstitial, micro-scale interrelations from which the order of social disciplining emerges. 'We all know the fascination which the love, or horror, of the state exercises today', writes Foucault in his essay on governmentality, but the state, he ventures, is not so coherent or self-contained; rather it 'is no more than a composite reality and a mythicized abstraction, whose importance is a lot more limited than many of us think' (1991: 103).

If the state is thereby out of focus in Foucault's politics, this is even truer of many theorists influenced by post-structuralism, more broadly conceived. To take just one illustration, in a text that is influential in anti-globalization, autonomist, and anarchist circles, John Holloway (2005) argues that successive generations of revolutionaries from Marx onward, have mistakenly and tragically assumed that to change the world they had to take state power. He indicts the left – largely the Marxist tradition – for an instrumentalist vision of the state, a fetishism in fact, but rather than provide an alternative definition, Holloway oscillates among several vaguely delineated conceptions of the state: the state is variously an empirical, indeed quite fetishized, grouping of institutions capable of being conquered (however wrongheaded such a strategy may be for Holloway); the state is a 'movement'; or it is simply a fiction. Thus he writes: 'the notion of citizenship is an element of the fiction upon which the existence of states, and particularly democratic states, is based' (2005: 96). The state here is at best chimerical; it is effectively dissolved as any kind of political target.

If this conceptual dissolution of state power marks Foucault and others off from Marx, it is worth emphasizing here that the inherent relationality of the state is shared by both thinkers, and indeed by Lenin, who saw the state as a concatenation of class relations: the 'state is a product and a manifestation of the *irreconcilability* of class antagonisms' (Lenin 1972: 9). Where they diverge is in their assessment of the power of the state. In a widely read essay, Timothy Mitchell (1999) registers the unsatisfactoriness of Foucault's formulation insofar as there is little sense how such a composite reality as the state is actually composed and of what it consists. He proposes instead that the state be seen as a historically developed 'effect of mundane processes' which not only give rise to the institutions commonly associated with state power, but also to the state's 'appearance as an abstraction' and its apparent (but not real) separateness from the society and the economy from within which the state draws its power (Mitchell 1999).

Yet insofar as these 'mundane processes' are not well specified (especially in relation to the 'state effects' on which Mitchell focuses) and seem to presume the rise of the specifically national state, paying little attention to expressions of state power at other scales or in pre-capitalist times, Mitchell's attempt to get beyond Foucault may not take us far enough. The insistence that we focus on 'state effects' while leaving the state and state power underspecified, guides our attention towards the objective albeit inchoate appearances of the state while leaving the historical, geographical and political economic substance of the state and state power quite out of focus. The latter is especially important: In its counter-fascination with a deliberately defocused state, this view cuts the threads of connection between the formation of the state and political economic relations. There is a deafening silence, in fact, about the pivotal role of the state in helping create the social and political conditions for capital accumulation, its abetment of the pursuit of profit. The state is effectively hollowed out, not just historically (if selectively) by its retreat in many places from certain social and regulatory commitments (if not others), but it is hollowed out conceptually and politically too.

Noting that anthropologists in recent years 'can neither leave the state alone nor decide whether it … has any … substance …', Gavin Smith (2010: 165) makes a parallel argument to the one I am pursuing here. The comparative vacuum of a loosely specified state in post-structuralist theory has its counterpart in the identification everywhere of 'state effects', going beyond the earlier project of identifying state effects in everyday life. The irony here is that far from displacing the supposed power of the state, the ubiquity of state effects raises the state to an almost metaphysical status, its own 'mythicized abstraction'; the state seems to be ubiquitous yet etherial, unable to be touched. The appearance of the state as recorded in its 'effects' thereby obscures any reality the state might actually have. Precisely at a time when the anti-globalization movement's promise of a world beyond (or at least oblique to) state power has dissolved in the acid of the techno-militarized surveillance state, this ubiquity of 'state effects' actually concedes considerable power to the state (or specific states), which are simultaneously displaced as targets for political action, political change. This seems to be a particularly dangerous knack, whether intended or otherwise, of our current political common sense.

This conclusion is not entirely new. More than a decade ago, Linda Weiss (1998) perceived that we had already entered 'a new era of "state denial".' In a parallel diagnosis, Tim Brennan identifies a 'poststructuralist common sense' (2006: 225), informing contemporary intellectual trends both in the academy and in popular discourse. The consequent 'striving for ambivalence as a matter of principle' and 'the ardent belief that answering a question forecloses it' (2006: 139) expresses the ethos of many contemporary treatments of the state. Indeed an appropriate motto for such trends might well be: 'see no state'. In this, poststructuralism is more complicit with, than critical of, the neoliberal ideological moment in which, of course, any regulatory state circumscription of capital accumulation is systematically disparaged and dismantled. The state is hollowed out in practice. It is not such a stretch of the imagination, nor is it at all perverse, to conclude that with its emphasis on the world-making power of discourse, post structuralism may in fact have worked, however unintentionally, as a fitting ideological garment for the performance of neoliberal capitalism.

So the dangerous knack of our current rhetoric may be an indulgence in a certain kind of fascination with appearance at the expense of meaning. This is particularly serious insofar as the contemporary state organizes against itself, becomes the champion of anti-state rhetoric. Self-defeating US military strategy in the early twenty-first century, from Afghanistan to Iraq to Guantanamo, surely highlights the actual contradiction of a supposedly hands-off superpower state. The current ideology of neoliberal governance craves shrinkage of the state, whereas state budgets continue to grow and state practice is transparently about consolidating maximal power. Far from being out of focus, a 'mythicized abstraction', or a concatenation of 'effects' more or less unhinged from social causes or processes, the twenty-first-century state is powerfully redefining its always contested prerogative through the broad lenses of security and terrorism.

This is not an abstract question. In August 2007, seven academic researchers and activists were taken into custody in Germany on terrorism charges under Section 129a anti-terrorist legislation. Urban sociologist and researcher Andrej Holm was singled out and imprisoned in solitary confinement for three weeks. Among the charges against Holm were: use of language, including the word 'gentrification', which the German state found suspicious and suggestive of terrorist connections; attendance at a meeting without his cell phone, which the state deemed evidence of conspiratorial behaviour. The urgency of again understanding the state, its power as well as its vulnerabilities, has never been greater.

Fascination and New Realities

'"Reality" dissimulates the world to us', Henri Lefebvre (2009 [1986]: 271) noted in the same passage in which he defines reality: 'the mundane ... serves as a screen' which 'resists becoming'. Hence 'the strange oscillation that ensures that the "real" seems wholly "unreal", like a dream and sometimes even a nightmare. And that the "more real", the world beyond this familiar reality, often seems surreal.' If the 'becoming' of reality is thereby masked, it is necessary Lefebvre says, 'to go beyond the mask. Into the world' (Lefebvre 2009 [1986]: 271). 'Fascination' can stick to the surface of what it sees, as in the self-conscious production of spectacle, thereby revealing the unreality of Lefebvre's 'real', but by inhabiting the dialectical crawl-space between appearance and reality, it might just as easily release the 'more real' in the form of the surreal. This is the central dilemma with 'fascination', particularly when it is dissociated from the social context, social struggle, and the contest of social interests that conspire to produce fascination.

All of this might seem a long way from our understanding of cities, especially in the context of the post-2007 global economic meltdown, a crisis largely blamed on prior decades of excess (economic and consumerist, militarist and cultural) at the top of the world's class hierarchy. The production of fascination in the landscapes of the Beijing Olympics (2008) or the Shanghai EXPO (2010), for example, are without doubt spectacular, from shiny new buildings to the more mundane violence of gentrification (an estimated 1 million were displaced in Beijing), and in many ways they bring together several divergent threads of the argument I am trying to make. First, these undertakings were explicitly intended as re-urbanization projects, and together with the globalization of the real estate development and financing industries as well as the equally global real estate crisis triggered by the collapse of the US mortgage market, they suggest the increasing centrality of city-building in the productive global economy. Second, there is a broad consensus that in the context of neoliberal globalism, some functions of the state are being rescaled, however partially and selectively, away from the national and either uploaded to the global or downloaded to the urban scale. Cities are becoming more important loci not only of capital accumulation but of service provision, social reproduction, and cultural identity. But third, insofar as the remaking of cities today is often organized around spectacular developments – malls, signature buildings by global architects, sweeping park and waterfront green spaces, self-proclaimed entertainment districts, and so forth – fascination is indeed enlisted to the cause of profitability and capital accumulation. Fourth, as the Chinese examples especially demonstrate, there is no contradiction between profit seeking and powerful direction by the state.

While the unfolding of the global economic crisis, euphemistically referred to as the 'Great Recession', may well halt some such developments, this is by no means certain. The Depression of the 1930s produced its own landscapes of fascination – dream or nightmare – whether in the retro-geological architecture of national park building in the US or the expansive social housing estates that came to dot the outskirts of many European cities. These were largely state-funded under contemporary Keynesian versions of today's stimulus plans. Today, the Chinese commitment to a massive stimulus-funded construction of infrastructure seems likely to extend this trend. On the cultural front, the 1930s saw inter alia a movie genre of flamboyant wealth and the emergence of urban noir suspense, while the science of the body, emblematized in the athletic, gymnastic body, moved from the laboratory into popular culture, not least through a fascination with the Olympics.

Economic and social crises, and responses to them, bring destitution for some but enchanting diversion for others; they bring a bit of both for many. In this context, I think we have to be very careful about how we approach the question of cities and fascination. Insofar as the political left has retreated in most places (but not Latin America) in the last three decades; insofar as a post-structuralist sensibility increasingly works as a new common sense, at least in many intellectual and cultural circles; and, related, insofar as political economic analyses have largely been displaced by new cultural lenses, the political left has already shown itself ill prepared to provide decisive analyses of the causes and direction of the crisis (or where such analyses have appeared the left has so far been unable to wrest the political initiative). What fascinates people in different places at different times is a vital window into that crawl space between appearance and reality, but we have to approach the question without succumbing to the idealist illusion – the most dangerous knack – that reality is a never-never land.

References

Anderson, J. 1973. 'Ideology in Geography: An Introduction'. *Antipode*, 5(3), 1–6.

Brennan, T. 2006. *Wars of Position: The Cultural Politics of Left and Right*. New York: Columbia University Press.

Castree, N. and Braun, B. 1998. 'The Construction of Nature and the Nature of Construction: analytical and political tools for building survivable futures', in *Remaking Reality: Nature at the Millenium*, edited by B. Braun and N. Castree. New York: Routledge, 3–42.

Elden, S. 2009. 'Reassessing Kant's Geography'. *Journal of Historical Geography*, 35(1), 3–25.

Feenberg, A. 2001. 'Marcuse and the Aestheticization of Technology', in *New Critical Theory: Essays on Liberation*, edited by W. Wilkerson and J. Paris. Lanham: Rowman & Littlefield, 135–54.

Foucault, M. 1991. 'Governmentality', in *The Foucault Effect: Studies in Governmentality*, edited by G. Burchell, C. Gordon and P. Miller. London: Harvester Wheatsheaf, 87–104.

Graham, S. (ed.) 2004. *Cities, War and Terrorism. Towards an Urban Geopolitics.* Oxford: Blackwell Publishers.

————— 2007a. 'War and the City'. *New Left Review*, 44(2), 121–32.

————— 2007b. *From Space to Street Corners: Global South Cities and US Military Technophilia.* Unpublished Paper: Department of Geography, University of Durham.

Holloway, J. 2005. *Change the World Without Taking Power.* London: Pluto Press.

Klein, N. 2007. *The Shock Doctrine: The Rise of Disaster.* London: Penguin.

Lefebvre, H. 2009 [1986]. *State, Space, World. Selected Essays*, edited by N. Brenner and S. Elden. Minneapolis: University of Minnesota Press.

Lenin, V. 1972. *The State and Revolution.* Moscow: Progress Publishers.

Mitchell, T. 1999. 'Society, Economy and the State Effect', in *State/Culture: State Formation After the Cultural Turn*, edited by G. Steinmetz. Ithaca, NY: Cornell University Press, 76–97.

Scahill, J. 2007. *Blackwater: The Rise of the World's Most Powerful Mercenary Army.* New York: Nation Books.

Smith, G. 2010. 'The State (Overstated)'. *Anthropologica*, 52, 165–72.

Smith, N. 1989. 'The Nature of Geography', in *Reflections on Richard Hartshorne's 'The Nature of Geography'*. Occasional Publication of the Association of American Geographers, vol. 1, edited by J.N. Entrikin and S.D. Brunn. Washington, D.C.: Association of American Geographers, 89–120.

————— 2000. 'Is a Critical Geopolitics Possible? Foucault, Class and the Vision thing'. *Political Geography*, 19(3), 365–71.

————— 2007. 'Disastrous Accumulation'. *South Atlantic Quarterly*, 106(4), 769–87.

Sokal, A. 1996. 'Transgressing the Boundaries: Toward a Transformative Hermeneutics of Quantum Gravity'. *Social Text*, 46–7, 217–52.

Weiss, L. 1998. *The Myth of the Powerless State.* Cambridge: Polity Press.

Weizman, E. 2007. *Hollow Land: Israel's Architecture of Occupation.* London, New York: Verso.

Chapter 7

The Urban Staging of Politics: Life Worlds, Aesthetics, and Planning – and an Example from Brazil

Wolf-Dietrich Sahr

Since David Harvey's famous *Social Justice and the City* (1973, partially revised in 2009) we have at our disposal at least one coherent theory of the city that develops a geography of value spheres. Following Karl Marx's footprints, Harvey interprets values as *relations* that originate from the social and material reality of things and human activities. It is common place knowledge that Karl Marx gave priority to two dimensions of values: use values and exchange values. Bearing in mind, however, the roots of his early writings this concept appears somehow reductionist, especially if we consider Marx's early interpretations of Hegel's idea of 'work'. Work is not an equivalent to labour, even if the German expression *Arbeit* does not allow this differentiation. Work (both noun and verb) is the expression of the self in its fullness (Marx 1990a [1844]: 574), and thus can be interpreted as a value of subjectivity (cf. Arendt 1958). However, at the latest since *Misère de la Philosophie* [*The Poverty of Philosophy*] (1990b [1847]), Marx restricted the term 'value' only to objective values, to values of corporal and functional needs, called use values, and to exchange values that are validated through corporal labour (!) offered in an open market. Such a dualism (or triad, if we include the subsequent surplus value) might be justifiable for theoretical reasons, but it poses serious problems when it comes to research on the material life world dimension, the subjective dimension. Consequently, in his revised edition, Harvey refrains from any dualist approach (2009: 16). Here, other value spheres appear to be equally important. Therefore, my argument investigates the existence, creation and use of more subjective values and validations that have gained relevance in postmodern geography. Specifically, I will target aesthetical values, and investigate their so-called 'superficiality' and appeal to emotions (fascination, attractiveness, rejection, repulsion).

This chapter, consequently, first reflects the theoretical impact of the division of space into differentiated value spheres, and then goes deeper into the question of relative and absolute values, focusing on dwelling and appearance. Afterwards, it turns to aesthetical concepts of form and atmosphere, and then highlights the idea of 'face value', an idea which appears laterally in the work of Guy Debord (1966), and literally in the term 'faciality' in Gilles Deleuze's and Felix Guattari's

A Thousand Plateaus (1988 [1980]). Finally, I will test these reflections in the case of Curitiba, a Brazilian State Capital in Southern Brazil, which for two decades has seen a governmental promotion strategy of urban marketing focusing on semiotic policies (see Müller 2004) while restructuring the city materially and 'incorporating' its population socially.

Not an Introduction, but an Immersion: the Fascination with Values

Since the beginning of an ideology of modernity in the nineteenth century, utilitarianism has been clearly revealed as a promoting force for modern society and its prime spatial form, urbanity. Karl Marx writes in *Das Kapital* [*The Capital*]: 'The utility of a thing transforms it into a use value' (2005 [1867]: 50).[1] Consequently, Marx agrees that in modern society 'utility' originates from a functional value system based on use relations that reconfigure space. In *Die Deutsche Ideologie* [*The German Ideology*] (1990 [1847]), Marx and Engels had linked such ideas to the functional socio-spatial arrangement of the city and the countryside. However, in *Das Kapital* (2005 [1867]) Marx admits that beyond any functional value of a thing there is also an inherent value: 'The commodity body itself, like iron, wheat, diamond etc, is therefore a use value or good. Its character does not depend on whether the appropriation of use characteristics causes high or low costs of labour for man' (2005 [1867]: 50f.).[2] Marx's remarks become clearer when we understand that for him labour is the decisive correlate of all market values, and thus becomes an absolute grounding of all social relations in modern society. However, continuing the aforementioned passage in *Das Kapital*, contradictions rapidly arise between absolute and relative values. Marx prefers to resolve them to the benefit of relative values: 'A use value or a good has only *one* value, because abstract human labour is reified or materialized in it.' (2005 [1867]: 53).[3] Thus, the absolute concrete element of work (a reminder of Hegel's self-externalization: *Selbstentäusserung*) turns out to become the relative use value of labour, and enters into a sphere of functional relations. Revoking this relative unilateralism, Marx then claims that all goods (as commodities) are embedded both into a 'natural form' and a 'value form' (2005 [1867]: 62), and thus introduces anew the difference of absolute (= natural) and relative connotations – with serious implications. Surprisingly, he now creates a microscopic uncertainty within his concept. A liminar vagueness appears at the transition zone between the expressions of *Form* [form] and *Wert* [value]. Both are now mingled into one, so

---

1   'Die Nützlichkeit eines Dings macht es zum Gebrauchswert.'

2   'Der Warenkörper selbst, wie Eisen, Weizen, Diamant usw., ist daher ein Gebrauchswert oder Gut. Dieser sein Charakter hängt nicht davon ab, ob die Aneignung seiner Gebrauchseigenschaften dem Menschen viel oder wenig Arbeit kostet.'

3   'Ein Gebrauchswert oder Gut hat also nur einen Wert, weil abstrakt menschliche Arbeit in ihm vergegenständlicht oder materialisiert ist.'

that Marx speaks of the *Wertform* [value form] (as an equivalent to use value), and its functional *Äquivalentform* [equivalent form] (thus denominating the exchange value) (2005 [1867]: 63). This seemingly minimal nebulosity between form and value, however, leads us to the edge of the concept of use value itself. Now, Karl Marx – eventually unintentionally – enters into a zone where values *appear* or *disappear* as forms, and not where they are construed through objective relations. Consequently, he arrives at this point at an area where the real is really a real 'body', a materiality beyond any relativization.

The moment of unrelated absolute values leaves us alone with 'mystery' and 'enigma', as Lefebvre says (1991 [1973]: 340). To be honest, I am not even sure whether it is not a *contradictio in adiecto* [contradiction in itself] to speak of 'absolute values', but I will use the term as a working expression. I suppose, an absolute value appears in a moment when semiotic relations of significance are philosophically interrupted, and nature (as natural reality) and commodities and fetishes (as artificial reality) become perceptible in their own form. Their proper absoluteness is different from their naturalization and commodification, processes that transform absolute values into relative ones when their categories are designed *in relation* to a background of non-rational and non-commercial worlds, exactly in the moment when they are coming into being as categories (*Gestalt*). In an absolute stance, relational thinking comes here to a provisional halt. Now, we have found the place where emotion and feeling can be perceived as non-relational rationalities of perception (see Hasse, in this volume). They arise in opposition to the prevailing rational market of relations based on a system of exchange values (Simmel 1996: 591).

The re-appearance of the sphere of non-relational absolute values can turn down any element of relation and, consequently, cause a systemic collapse (entropy) in a relational world. This seems to be the case in actual postmodern capitalism. Here, the main social danger arises from the veneration of the overwhelming presence of capitalism (both empirically, but also expressed in all critical stances towards capitalism): what is at stake in capitalism is the danger of an absolute exchangeability of social life, its complete dissolution into a melting pot of relative values, the end of subjectivity. However, we are not in a situation of despair. Instead, there is a chance what will eventually free us from the prison of relational thinking (and its corresponding mode of rationalized capitalist production), proposing new values, social values of a post-capitalist society through subjectivity. Here, we can expect space for a new 'reality', a reality of 'position', or 'opposition', a reality that can be understood as a rock-like location situated beyond the veil of network relations (including functional, social, and exchange relations).

In this sense, it might be possible to abandon the clear divide of a 'sign' into a signifier and its content. Usually, a sign is embedded in a network of relations of signifiers, and its absolute 'reality' appears in its content. Some authors, mainly influenced by semiotics, have insisted on this socio-semiotic divide, making it the basis of logical thinking (e.g. Gottdiener 1995: 27). Overcoming such dichotomies, however, permits us to develop a new cartography of things, signs and subjectivities altogether (Guattari 1995 [1992]) pleading for new forms of the territorialization of subjects (Deleuze and Guattari 1988 [1980]). Such an approach is not at all a rejection of logical differentiation, but it is a specific disapproval of a strict separation between the sign and its underground, between consciousness and the subconscious, between the real and the surreal. This diffuse conception leads to the investigation of another 'reality', a reality that is closer to fullness and experience, located far beyond the simple rationalistic 'real'.

So when I now question, and at the same time procure, 'reality', it is not to deny or to confirm reality in one way or the other, searching for a 'better' truth with a better methodology, but it is to find a place, an anchor point in a world that is interpreted as a network of relations (values) between absolute material contents (persons, goods etc.). Baudrillard speaks out on the actual panic of academics who perceive that their values have 'become dissociated from its content and begin to function alone, to its very form' (Baudrillard 2001: 153). Such panic seems to be understandable when we consider that our thinking and perception are no longer grounded in 'reality'. Baudrillard's essay *Values' Last Tango* truly reveals a 'universe of simulation' that is 'transreal and transfinite: no test of reality will come to put it to an end' (ibid.: 155). But in my opinion, such a philosophical attitude does not call for a fatalistic attitude. Instead, it opens the way for a possibilistic concept of renewed interrelations between different spheres of truth, each one with its own relative absoluteness in a specific social context.

The actual processes of spectacularization and commodification in postmodern cities seem to reveal such a situation in a specific way. In urban everyday worlds, most inhabitants have encapsulated their lives individually into a city of 'homes'. But they are still dominated by old-fashioned values of functionalism (utilitarianism), exchangeability and their respective network structures. Therefore, psychological, aesthetical and emotional attitudes arise in opposition to such unlimited relativeness.

Postmodernization clearly targets citizens as consumers via these attitudes. It offers a new world of simulacrum and appearance, and thus subverts both the traditional functional relations of use values (in their utilitarian and social sense), but also the hidden social relations of commodities within exchange values. In fact, the 'rational' distance between a good and its value shrinks to a new trinity of appeal, impetus, and spontaneous reaction, forming new energetics in a kind of 'formism'. According to Maffesoli, formism represents a relevant structuration principle in postmodern societies (see 1990, 2007). It includes the collapse of distance between the subject and the object, and implodes the 'psychological relation between us and the things', as Georg Simmel put it in his *Philosophie des Geldes* [*Philosophy of Money*] early in the twentieth century (cf. Simmel 1996: 73). Thus, new value categories of a more psychological nature become visible in a *raison sensible* (Maffesoli 1996), and a new social analysis now has to contrast with common social theory, which still prefers to describe social situations in relational terms: consequently, most critical authors continue to refrain from profound investigation of psychology, interpreting the collapse of rational relations in capitalism only as a strategic treachery initiated by capitalist agents to target consumers as superficial beings. I shall attempt, however, to argue that the success of such strategies is only comprehensible when we go beyond the superficial interpretation and investigate the serious changes that are effected in the abyss of subjectivity and social life. Such a sphere is far beyond the reach of simple capitalist agents, and also beyond the imagination of simplistic critics of capitalism.

In the context of unveiling sociological truth in favour of social truth, it becomes necessary to look into immediacy and body, emotions and feeling, touch and fetishism in a positive sense. That does mean accepting these elements primarily as social situations that have survived from the rational 'collapse'. This point of restart now enables us to adopt a different kind of critique: while superficial postmodernism refers mostly to floating signifiers, and while materialists and positivists still insist on the abstraction of basic needs and exchange values, the new concepts, mostly of a poetic character and based on profound imaginations (not simple images and sceneries), induce new sensitivities that can bring us closer to the edge of conventional academic concepts, and revalidate aesthetics (Welsch 1990). Now, the life world spheres are dealt with in a different critical stance: the critique is directed towards the relation between signification and content, and consequently paves the way for sociological concepts different from semiotics. Some authors call it non-representational geography (Thrift 1996, 2007, cf. also Anderson and Harrison 2010, Lorimer 2005). However, my argument is going in a slightly different direction. It ventures into the aesthetics of social geography: a geography of phenomenological dwelling, of forming, and of appearance.

Dwelling and Forming: Cutting Out a Life World Form from the Relational World

In *The Urban Revolution* (2003 [1970]) and in *The Production of Space* (1991 [1973]), Henri Lefebvre made several references to two famous essays of Martin Heidegger. Both deal with the question of *Wohnen* [dwelling]. *Wohnen* is a profound and very emotive German expression that appears in both titles: *Poetically Man Dwells* (Heidegger 2001a [1951]: 209ff.) and *Building, Dwelling, Thinking* (Heidegger 2001b [1951]: 141ff.). Basically, Heidegger's essays tackle the question of dwelling in a deep philosophical sense. With this somehow mythically connoted expression, Heidegger tries to overcome the traditional semiotic separation between form and content. In *The Urban Revolution* (2003 [1970]), Lefebvre follows Heidegger and characterizes dwelling – in his chapters on the urban phenomenon (Chapter 3) and on its spatial levels and dimensions (Chapter 4) – as a 'form', which gains its concreteness (it may be fully expressed or degraded) in a sphere where the unconscious and the conscious mingle into a full life. Such life is full when it is fulfilled poetically. This was the meaning of the words of the eighteenth century German poet Friedrich Hölderlin: *Poetically Man Dwells*. Hölderlin's poem had inspired Heidegger to interpret poetry as a social force that creates a space for ontological habitation. Thus, poetry is not simply a branch of art, but an action, a performance introducing a 'poetics of space' (Bachelard 1994). Lefebvre (2003 [1970]) follows a similar track when he favours the investigation of the urban element beyond any decodification of urban reality, and instead focuses on *habitare* as a way of living. At this moment, unlimited linguistic, scientific, and political modern abstraction, which are based on relations between signs and content, have been put aside to be supplanted by profound phenomenological, experienced, and aesthetical concepts of form. These go far beyond the simple form-content differentiation of Hjelmslev (1971). Consequently, Heidegger's essays represent philosophical reflections that are located on the edge of classical socio-semiotics. Poetry and art, probably more than any social science, open a gateway for grounding 'sociality', in contrast to the relational world of use values and exchange values as 'sociabilities' (Maffesoli 1996). In *The Production of Space* (1991 [1973]), Lefebvre accuses Heidegger of thus proposing an ontology of absolute space that deflects the focus of attention from the temporal (and relational) aspect of production (ibid.: 122). Lefebvre might be right, but his statement shows that his own position does not really consider the full revolutionary potential of the 'resistancibility' (*Widerständigkeit* in Heidegger's words) of dwelling.

For Heidegger, every form is the separation of a thing from its environment. Simultaneously, every thing is embedded into the *Wertprädikate* [predicates of values] via form (Heidegger 1993 [1926]: 99). Thus, forming is a transition zone that grounds each thing ontologically into being. Heidegger suggests that the interior of a being and the surrounding environment should be bound together in space: then they are spatially related in distance and proximity and form an *absolute* relation, a specific organic combination of elements and lines (ibid.: 101–105, Deleuze and Guattari probably would call it 'territorialization').

Here, art comes into play. Art is a specific trait of social being, aside of semiotics, function, market, and society. Its aesthetics can be interpreted as an expression of social truth (Heidegger 2005 [1935]: 30). Following Heidegger, the social truth of art is achieved through the mediation between things (ibid.: 13) and works (ibid.: 19). In a moment when things serve a purpose, Heidegger calls them purposeful things, in German *Zeug*. *Zeug* is now seen as a mean of mediation, a specific equipment of revelation of humanism between ontological being and relational utility (ibid.: 21). The ambivalence of such mediation becomes clear when Heidegger refers to the spatialization process of 'clearing'. Clearing (in German the verb is linked to *räumen* [spacing]) is understood as opening a space, for example, in a forest. The metaphorical clearance, the creation of a new space of meaning in its own environment, is a piece of art (ibid.: 51). Here, the *Gestalt* (ibid.: 67) is the result of cutting off relations with its surroundings, just the opposite of what capitalism tries to do via economics and what semiotics and sociology propagate via their theories. Such an artificial and artistic process of 'cutting out' evacuates the piece of art from the networks of relative values. Instead, it reveals it in itself, in its autonomy. Within this reserved space, however, we do not assist solely a separation, but we also perceive a transformation of values.

Now the habitual relations of the *Welt* [world] as ambience, and the relations to the *Erde* [earth] as natural groundings involve a new *Tun und Schätzen, Kennen und Blicken* [doing and evaluating, knowing and viewing] (Heidegger 2005 [1935]: 67). This kind of revolutionary aesthetics is magic. It can be called fetishism, but this fetishism is not bound internally through immediate attraction and binding, as it is in consuming processes in late capitalism; it is the result of singling out a piece of art aesthetically, it develops its own ontology with a specific real world character – different from the outside space, but liable to revolutionary identification.

Heidegger's anti-environmental approach to space found a critical parallel in Bakhtin's essay on *The Problem of Content, Material and Form in Verbal Art* (Bakhtin 1990, Bachtin 1979). Bakhtin demarcates the aesthetic operation as a delimitation process in the unlimited open fields of knowledge (gnoseology) and agency (ethics). Purposely, he does not link aesthetical values to the relational value systems of the outside (use values, exchange values, social values, and semiotics), but cuts them off, so that aesthetics can develop their own value (Bachtin 1979: 95). Bakhtin clearly admits that this understanding of art refers to the autonomy of art; but for him it is exactly this autonomy which characterizes the critical (and relational) potential of art to (en-)counter the outside world. While each artist induces a specific space into the general social atmosphere, he develops contingent values that react – in a critical stance – to the events of the social fields of knowledge and agency (and their respective value spheres). Citing Sloterdijks *Abrüstung [Disarmament]* (1986), van Tuinen refers to such a situation as a *Konvergenz von Kritik und Ästhetik* [convergence between critique and aesthetics] (2006: 47).

Joseph Beuys once said: 'Everybody is an artist' (quoted after Adriani, Konnertz and Thomas 1981: 300).[4] Thus, he pointed to the fact that dwellers (and we include especially urban dwellers) have their own artistic 'techniques' (*techné* is the Greek denomination for art) to create a 'home' for dwelling. According to Beuys, everybody is endowed with the ability to counter the overwhelming network of exchange relations, whether he or she is treated as a 'consumer' or not. So the investigation of how autonomous space can be created through aesthetics within urban life worlds, and how its respective values can be developed, is of prime importance The Brazilian geographer Marcelo Lopes de Souza has clearly highlighted the necessity for a socio-geographical *projeto de autonomia* [project of autonomy] (2006: 68), proposing a different epistemological attitude towards capitalism than classical criticism, and makes his stance for new politics, ethics, and even (inconclusive) aesthetics of autonomy (ibid.: 88). In empirical investigations in Latin American cities, we can learn how urban dwellers react to the psychological (and superficial) attacks of an entropic world market system, while preserving their own subjectivities and forming their own Greek *agoras* amid a sea of conformism and relativiziation. 'Formism' then is, in its profound sense, a prime means of spatial resistance (Maffesoli 1990).

---

4 'Jeder Mensch ist ein Künstler.'

Thus, the social technique of formism involves largely non-rational attitudes, such as the creation of 'magic' spaces that link nature and commodities with man, leisure and artistic atmospheres. This includes a revaluation of atmospheric feelings, like fascination or fear, and the whole field of psychedelics. Such focus does not impede the recognition of erratic elements of rationalism in the life world, but it procures deeper forms of living, forms that are creating atmospheres, so that the aesthetic process of 'forming' itself and the psychological process of 'sphering' appear in an intertwined constellation.

Atmospheres and Fascination: Aesthetics (Un-)limited

Some authors in German speaking geography (Kazig 2007, Hasse in this volume) have recently called attention to the atmospheric character of life worlds (see also Bærenholdt, Haldrup, Larsen and Urry 2004 for Danish geography). They recall the attempts of some philosophers and theorists of art and aesthetics at the end of the 1990s who also referred to concepts of atmosphere and performance (Fischer-Lichte 2004, Seel 2000, 2007, Sloterdijk 1998, Böhme 1995). Even more recently, the 'atmospheric' work of Peter Sloterdijk has received attention in the English-speaking geography. So Elden and Mendieta (2009) organized a special issue on Peter Sloterdijks 'geography' in *Environment and Planning* (see also Castro 2009, Morin 2009), but unfortunately most invited authors to that special issue have remained curiously tied to a traditional 'critical' stance, possibly due to the still scarce availability of the immense oeuvre of the German philosopher in English.

The atmospheric approach is characterized by an attempt to dissolve the customary differentiation between form and content, and instead favours a medial position of performance that rethinks the connection between man and environment. Grounded in the philosophy of the *Neue Phänomenologie* [*New Phenomenology*] of Hermann Schmitz (2003, 2009), the aforementioned German authors characterize atmospheres as *Halbdinge* [half-things], which are not constant but sporadic. Atmospheres, for them, are limited space-time events. Therefore, they cannot be seized by subjects as objects, but they themselves involve and surround the subjects as objects (Kazig 2007). Here, the emotional component comes into play. Emotions and feelings are not directed, as corporal action is; they are diffuse, like an atmosphere, and call for a specific constellation of subjective sensitivity. In this situation, it is analytically impossible to differentiate between what half-things are *objectively* and how they are perceived/interpreted *subjectively*.

According to German phenomenologist Bernhard Waldenfels, attention (a near 'relative' of fascination) is an act of becoming conscious in a medial situation: something appears to our mind while we are simultaneously creating the object through our perception (apperception) (Waldenfels 2004: 66). This 'active' appearance of things is characterized by both a spatial and a temporal aspect (Seel 2000: 44). Thus, it results in a situation where attention is paid to something through (con-)centration, through encircling. Such a performance neither inclines to the objective side, nor simply focuses on the subjective side, but remains located in-between. In this context we can speak of a situation of true encounter, where resistible things come to our mind without rational and reflexive distancing. Leaving behind the rationalizing bridge, attention thus provokes feelings – they may be fascination and/or fear, attraction and/or repulsion. Their non-rational relations are construed via the psychological process of focusing, condensing, and through the delimitation interpreted in aesthetic processes.

Peter Sloterdijk comprehends territories beyond the subject–object dichotomy as a kind of psychological geography, where the interior world is turned to the exterior but also torn from the exterior, resulting in crystallized atmospheric islands (2004: 338). For him, such a 'worlding' causes a specific topology, and the condensed atmospheres of the interior are expressed in exterior glass house architecture. Such geography is the result of *Raumbildung und Raumatmosphärenkontrolle* [space creation and the control of spatial atmospheres] (ibid.: 346), truly geographical principles that are valid for engineering, psychotherapy, politics, and others (see also Sahr 2009). Sloterdijk has presented these ideas in his monumental oeuvre of *Sphären* [*Spheres*] (1998, 1999, 2004), where he explains the actual society as a divide between two processes of sphering: globalization, which is based on the unlimited networks of exchange values, mostly prevalent in macrospherology (Sloterdijk 1998, 2005), and globulization (immunization) as a basic principle of microspherology and subjectivity (Sloterdijk 1999, 2005). Both globalization and globulization, however, are still bound to a third dimension: the creation of *Schäume* [*Spume*], which is a kind of double bind in a *Plurale Sphärologie* [*Plural Spherology*] (2004).

Following the propositions of Sloterdijk, it now becomes clear that atmospheres are not necessarily unlimited. They are geographically focused on emotional attitudes – and fascination, attraction, fear, and terror are among them (Böhme 1995: 63). Thus, atmospheric aesthetic values appear visibly as emotions in our faces. And so we turn our discussion finally to a phenomenological aesthetical value, the 'face value'.

The Face Value: a New Superficiality, or a Glance into the Abyss of Emotion?

It is a basic character of atmospheres that they do not have a specific reference. They are limited temporarily and spatially, relating perceiver and environment to one another via diffuse feelings/emotions. Unfortunately, we still often interpret emotion in the perspective of modernist psychology. Thus, we link it to individuality and subjectivity (it even seems that 'feelings' have become the last resort of individuality in modern anthropology). But the actual performance of emotion goes beyond this. Emotions are based on immediacy and collectivity. They avoid reflection, but enable 'location' both in a transitive (= position of ...) and an intransitive form (= locale, environment). So the perceiving socius – be it an individual, a group, a society, or a mass – is largely *moved* by e-motion, and appeals and impulses constitute 'truths' of their own. So it is no longer possible to speak of an overall social truth, as was the case with use values in historical materialism, or to refer to religious or philosophical truth, as some dogmas of religions propose. The globalized atmosphere of the emotional truth is limited, but appears as an absolute value in itself. Such an atmosphere comes up erratically, cutting the *globulus* from the surrounding networks, and it appears especially at moments (as in actual postmodernism) when the social and economic configurations of relative values cannot be secured, or are in crisis.

In this context, an essay published by Guy Debord in *The Situationist International* of 1966 is very interesting. Debord – commenting on the race riots of 1965 in Los Angeles – describes the looting of supermarket shelves by a furious crowd. He points to the fact that most people who entered the supermarkets simply grasped what came into their hands, without really searching for specific goods that could satisfy their needs. Debord even makes the remark that, at that moment, the goods on the shelves had shown their 'face value': they looked at the looters, and thus their commodity values (use value and exchange value) had lost sovereignty while their appeal provoked the immediate desire of the looters. Thus, the angry crowd had overcome the principles and rules of capitalist organization by revealing the profound human reality of things. The objects of consumption had become immediate objects for the subjects. They called for attention, while their semiotics had collapsed.

In a similar vein, Böhme criticizes a classical oeuvre of consumer aesthetics, Wolfgang Haug's *Critique of Commodity Aesthetics* (1986 [1971]). While Haug interprets modern consumer capitalism as a strategy for stimulating commercial capitalism via visual techniques, creating promotions and labels for goods, Böhme understands the present consumer society as a social body, where the original 'needs' (whether satisfied or not) are accompanied by a psychological structure that differs from the need structure (1995: 64). For him, society is designed much more through desire. While needs can be satisfied, desires cannot; but while desires can be created artificially, needs are corporally (naturally) bound. The question embedded here, finally, is whether we understand society as a 'natural' or an 'artificial' body. Böhme assumes the second alternative and claims: 'Developed capitalism has become an aesthetic economy' (1995: 64).[5]

The development of aesthetic processes in such a society includes the creation of spatial atmospheres, a position that reinforces the argument for a third value sphere in addition to the two prime value spheres of Karl Marx. While use values can be interpreted as pre-semiotic or proto-semiotic, due to their corporal need structure embedded in denotative functions and naturalization, exchange values are semiotic structures that depend on a clear separation between content and sign (in capitalist practice transformed into the difference of commodity and money). In late modern capitalism, the exchange sphere is so amplified by wildly floating

---

5   'So wurde entfalteter Kapitalismus zur ästhetischen Ökonomie.'

signifiers that the difference between goods and values implodes, permitting goods to now be fully manipulated with superficial aesthetical creations. These postmodern goods originate from aesthetical processes beyond need values, so that semiotic connotation supersedes denotative function. However, aesthetic values themselves are more than connotations. They are somehow post-semiotic, as their sign-content structure is dissolved through diffusion, substituting nebulous and cloudy agglomerations for clear forms. Parameters of spacing and combination now gain a new validity. They introduce us to a new (old?) value category, the mystical sphere. Consequently, the sense of meaning is waning, while the sense for aesthetics is gaining space in event-moments.

Deleuze and Guattari (1988 [1980]) have named event-moments, where subjectivity and sociality are performed, 'agencement'. Unfortunately, the English translation of the French *agencement* in *A Thousand Plateaus* is 'assemblage', a translation that considerably inhibits its connotative understanding for creative activity. The new ethno-aesthetic paradigm of *agencement* received its specification in *Chaosmosis* (Guattari 1995 [1992]), where it is divided into three sub-dimensions. The first *agencement* is proto-aesthetic, where subjectivity (identity, collective thinking) finds specific forms of expression in a plural map of territorializing the self and others (ibid.: 101). The second *agencement* is built upon absolute semiotism. It refers to generalities like the 'truth' of logical idealities, the 'good' of moral desires, the 'law' of public space, the 'capital' of economic exchange, the 'beauty' of aesthetics (ibid.: 103). Such valorizations are based on (invented?) instances such as God, Being, Society, Significance etc. Here, subjectivities lose their (interior) individuality, and are now driven by exterior values. The overall presence of values of this kind (for example, money values) creates a field of relative deterritorialization. The third *agencement*, however, the processual *agencement* is more complex. It depends fully on aesthetics (perception, affection) and works as a kind of auto-poetic machine (ibid.: 105). This machine is permanently reorganizing the social (ethical) and scientific (knowledge) fields, using absolute deterritorializations. Now the spheres (Guattari speaks of 'universes') of value are simply crystallized or decrystallized. They are spheres of endless creativity (power, energy), while they appear in sensitive finitude (aesthetic forms). This flexible understanding of territory is clearly geographical. The Brazilian geographer Rogério Haesbaert (2004), who has closely investigated the interrelations between the Deleuzian-Guattarian approach and contemporary geography, explains that the permanent process of deterritorialization, reterritorialization and multiplicity invests in a system of multiterritoriality. Thus, the combined structure of 'subject and sign' seems to be one of the decisive elements of a geographical structuration of spatial forms.

In *A Thousand Plateaus* Deleuze and Guattari (1988 [1980]) define the combination of subjectivity and significance as 'faciality', referring to a kind of post-semiotic situating. A 'face' reunites the interior (the subject in its specific subjectivity) with the exterior (signs) in a corporal, visual and atmospheric expression (ibid.: 189), similar to a landscape (ibid.: 191). What we called dwelling before, has now received a 'face'. The face is different from a sign: its expression is not only superficial, but externalizes the interiority of man, and vice-versa. Facialization then is more than a mere semiotic fact: it confuses emotion and expression (as non-rationalized elements) with social binding and empathy, resulting in new forms of 'atmospheric' (agglomerative) socialities. In this moment, a face value overcomes the predominance of exchange values, but also the plurality of use values, and turns our environment, e.g. the city, into a map of social atmospheres. Therefore, spatial processes like urban construction, urban planning, and urban living can be interpreted as facializations of space.

From this perspective, I would like to illustrate (and not analyze) my reflections briefly with some facial expressions of Curitiba. Curitiba is a Brazilian city which, since the 1970s, has focussed on semiotic techniques in its urban planning process (Müller 2004), being one of the pioneers in Latin America in using face values for the emotional structuration of a city.

The Face of Curitiba (Brazil): a Home for its Citizens?

Curitiba is a special city in the context of Brazil. Since the 1970s, it has acquired the fame of being well-administered, of having an excellent transport system, being preoccupied with historical and natural conservation, as well as looking after its citizens – even praised for the excellent work of its municipal Planning Institute (*Instituto de Pesquisa e Planejamento Urbano de Curitiba* – IPPUC, founded in 1965). In this context, some scientists have investigated the semiotic and discursive processes of its city marketing from a critical perspective (e.g. Caetano, Duarte and Ferrara 2007, Irazábal 2005). Others demystify the city as a *Spectacle City* (Sánchez García 1997), *Model City* (de Oliveira 2000), *A Reinvenção das Cidades para um Mercado Mundial* [*City Invented for the World Market*] (Sánchez 2003), or simply as a *A Arquitetura do Desejo* [*City of Desire*] (Dias 2006), usually pointing to the immense incongruence between the city's 'reality' and its propagandistic 'appearance'. A third group of researchers easily identifies the failures of Curitiba's metropolization process when demonstrating its considerable economic and social disparities, visible mostly in quantitative and structural data (Moura and Firkowski 2009, Zirkl 2007). However, nearly nobody refers to the socio-psychological and aesthetical effects that are produced in the city via architectural insertions.

Thus, most authors still follow the classical epistemological dichotomy between a material-corporal 'perceived space', lived by the population in spatial practices (use value), and an idealized 'conceived space' created by 'scientists, planners, urbanists, technocrats' as 'representations of space' administered to 'capitalize' on surplus values in a system of exchange values (Lefebvre 1991 [1973]: 38f.). When it comes to social aesthetics, however, these approaches fall short. They omit what Henri Lefebvre has called the 'representational space', where urban dwellers develop their subjectivities within a framework of a physical structure, in a kind of architectural *Globulisierung* [globulization] (Sloterdijk 1999). In this respect, the 'facial' expressions of the city can be interpreted as a conglomerate of subjectivities. These subjectivities vary and are divided among administrative subjects, citizens, dwellers, consumers, ethnic groups, social classes, or even simple locals, all represented in a gigantic map of a socio-architectural *agencement*.

Curitiba has a considerable number of architectural insertions based on specific face values. Among them are the *obras* [works] of the municipal government, a large number of churches, the buildings of social institutions and societies, like football stadiums, clubs, private museums, and a considerable number of monuments of the commercial sector, like shopping malls, galleries, and pedestrian zones (Sahr 2001). The 'environment' of these insertions often exceeds their physical location, and they develop an aesthetical and atmospheric centrality. Thus, architecture binds local identities in representational spaces, in a technique of a 'semiotic baroque' (Irazábal 2005: 274, see also Chapter 5 in Maffesoli 1996) that produces semiotics and subjectivities simultaneously. In contrast to Lefebvre, who insists that most representational spaces reveal habitus-like attitudes grounded in the dwellers' personalities (1991 [1973]: 246), the approach of 'face values' demonstrates that the construction of urban subjectivities is continuous and mostly observable in architecture; it literally takes place on the urban surface. Due to the atmospheric paradigm, such a situation is not relational, but positional. Its atmosphere is bivocal and mediated between the hierarchical actors (government, churches, private investors) and the urban dwellers, and induces an aesthetic economy of attention and fascination in specific locations (see Schmid 2009, Franck 1998).

Curitiba has been the capital of Paraná State since 1853. It has a relatively long planning history, with some urbanist interventions being observed as early as in the late nineteenth century. This is the case of the first City Park in 1885, the installation of a railway quarter around the central station at the beginning of the twentieth century, and the implementation of a well-equipped tramway system at the same time; all interventions transported the values of positivist progress for the local population. The city's first Master Plan was designed in 1943. It had been developed by the well-known French urbanist Donat Alfred Agache (Menezes 1996: 64) and divided the city into different functional centres (business, civic, politechnical-universitarian, industrial, commercial), all of them with some specific architectural highlights that become facial expressions of the city. These centres were interlinked through a rational transport system (de Oliveira 2000: 74ff.). However, in the 1940s the city's fame due to this plan was restricted to some urbanists, as it was only in 1971, during the military regime, that the plan – now modified – was implemented physically by the planning institute IPPUC.

It is specifically due to this Institute that the city has gained its reputation as a model city. Though inaugurated with political intentions, the IPPUC is a technical advisory unit, only indirectly linked to the administrative structure of the city (*Prefeitura*). In IPPUC's view, Curitiban dwellers are mostly seen as administrative objects, and not so much as citizens that actively participate in politics. However, IPPUC's political influence is considerable. Since 1971, first through dictatorial nomination and then by plebiscitarian democracy, nearly all the mayors of the city have been somehow connected to the Institute. It seems that the population has profoundly identified with IPPUC's urban policies, in spite of often uttered dissatisfaction with the infra-structural deficiencies of the city. Such a psychological link between planning, politics, and the perceptions of the population creates a 'stability' of the urban surface, which is not very healthy in democratic terms, but gives room for the implementation of long term urban subjectivities. Thus, the actual city is much more configured through aesthetics and facial expressions than through a diverse field of social conflicts, or through being a product of a contradictory capitalist structuration processes (Delgado and Deschamps 2009).

Such an atmospheric cohesion is the result of the efficiency of an aesthetical policy. In this sense, Curitiba was coined a 'human city' by Mayor Jaime Lerner in his inaugural speech in 1971, when he identified the city with a 'soul', a vital force which permeates all citizens. In Lerner's perspective, the city had to be at the service of man and man did not have to be subordinated to the city as a spectator (cf. Menezes 1996: 94-95). These words, uttered by a mayor in the most violent period of Brazilian military government, demonstrated how the municipal government of that time substituted social interaction (democracy) with administrative aesthetics. It made urban inhabitants thus identify with their environment through conservationist policies both for natural and historical sceneries, simulating peace in a profoundly disrupted society, and comforted urban dwellers in their difficult everyday lives by looking after its subjects. In his third term, now democratically legitimated, Lerner insisted on such an aesthetic approach, and most of his successors followed this policy. So, until today democratization has never really reached the facial expression of Curitiba, and there is no place on the urban map for social controversies and disputes. While social segregation is even reinforced through urban planning, meeting points are rare, and are mostly relegated to activities of the (weak) state, where they become emptied from their common social value due to the technical approach of the city government. Several 'successful' urban projects reveal this fact. In what follows, I will briefly analyze the construction of Curitiban subjectivities through the highly praised bus system, the citizen street programme, the lighthouses of knowledge, the city parks, and, as the only intervention of the private sector, the diverse map of shopping malls.

The first municipal government of Jaime Lerner was characterized by the introduction of a bus system known world-wide. At that time, bus stops on the main traffic routes were equipped with transparent tubes, where the passengers had to wait for the vehicles while protected from rain and wind. Such a configuration allows an organized and regulated transportation system (use value), makes an economic approach viable (exchange value), and the passengers visible (face value). They thus represent the ideal of 'freely moving' citizens as a postmodern expression. Therefore, it is understandable that most bus tubes can be found in areas of bright insight, like avenues, squares, and in front of public buildings and private commercial enterprises. Here, the citizens are exposed in self-referential aesthetics. But in spite of some technical critiques, most users publicly identify with these installations and highlight their advantages, especially when they speak, full of pride, to foreigners. Thus, the tubes have become facial marks in the urban landscape, and demonstrate a specific functional identity that integrates the life worlds of the dwellers with the image of their city.

Another example of a governmental urban aesthetics is the great number of bus transit centres on the main lines. In some places, these transit centres are connected to so-called 'Citizen Streets'. The 'Citizen Street Programme' came into being in 1995, and established local civic centres, where popular administrative functions are available in spaces of a specific architecture. These buildings are open galleries, painted exclusively in yellow, orange, and red. Again, the centres are plainly visible and exposed, often attached to other activities, like public popular shopping malls, churches, or museums. Thus, the citizen streets represent a similar conception of face value, exposing administered urban subjects as in the bus system.

A third element implemented in the 1990s – in a period of increased concern with the education system – are the so-called *Farol de saber* [Lighthouses of knowledge]. These are educational centres where libraries and computers are accessible free of charge. The lighthouses have been installed in great number all over the city and are well accepted among the local population, especially in the marginalized sectors of the city. Again, their layout is homogenized, demonstrating their socially integrative effects. Like the citizen streets they are painted in appealing colours; furthermore, their structure represents a tower with a globe on its top connotating the *Lighthouse of Alexandria*, where the most famous library of Antiquity was located. Such a connotation clearly adheres to the bourgeois understanding of education, furnishing unidirectional knowledge instead of bidirectional reflection.

A fourth element of facial planning is the installation of green spaces and public thematic parks. The parks theme specific cultural expressions (like ethnicity, leisure, or specific knowledge, e.g. the Botanical Garden). Most of the parks stretch along urban rivers, thus preventing their occupation through illegal settlements, but also increasing land speculation in the annexed neighbourhoods. Though distributed unevenly over the city (with dominance in the richer sectors), the parks are well accepted among large portions of the population, especially during the weekends, when they receive enormous crowds. Then, even a certain social mix can be observed, a fact that is relatively rare in Latin American societies. In this respect, the parks have become instruments of facial pacification mediating social tensions via a natural, or rather a pseudo-natural, and romantic atmosphere.

A fifth element of Curitiba's facial expression is the large number of shopping malls and popular shopping areas in Curitiba (see Gil 2003, Sahr 2001). Shopping malls are well distributed all over the city, and they are differentiated for different social classes. Thus, they attract their consumers and visitors by specific atmospheres. There are shopping areas of representation, of functionality, of popularity etc. Some are installed in traditional sites of the city (like an old factory, the central railway station, or some military barracks) appealing mostly to the traditional conservative middle class, while others expose their aesthetics of postmodern spaces of glamour and brilliance to elitist social groups. Some reproduce the buzzing atmospheres of popular market halls, while others target the specific aesthetics of urban tribes. From such a perspective, the shopping malls are centres of irradiation for social aesthetics, and reproduce the urban space within diverse consumer subjectivities.

All these examples of urban facial expressions are moulding urban space aesthetically. Their limited atmospheres are embedded and highlighted amidst areas of less atmospheric significance, areas that are often polluted by a confusing architecture and disordered urbanism. Thus, atmosphering the city represents a planning strategy of postmodernity. It focuses on specific portions of the urban space, and here induces fascination by responding to the emotional subjectivities of its inhabitants. In contrast to strictly planned cities, which are construed on the principles of rationality and functionality (avoiding a free emotionality), the vague urban planning process in Latin America has to seek refuge in emotional factors to reinforce the city's internal structures. Thus, it responds adequately to the partially amorphous social corpus of Latin American societies themselves. In such societies, atmospheres are more important than rationalization. They become a socially structuring factor. Thus, fascination might be evaluated by some critical scholars as a kind of deceiving situation, as it simulates social coherence where it is not perceptible rationally, but on the subjective level the emotional moulding is a social necessity that stimulates social coherence. Therefore, fascination should not be underestimated, since an aesthetical framing also has its critical sense.

Conclusion

This fragmentary investigation of the facial expressions of Curitiba reveals how aesthetic social framing works, and where the proposed facial value approach exceeds the analysis of power relations. While power analysis represents urban dwellers in abstract and objectified relations as knots and dots within networks and streams of power, even if some restricted or imposed forms of subjectivity are allowed, the investigation of aesthetic forms demonstrates the contradictory opening and closing of subjective territories via autonomous creativity and energy. Hence individuals and social groups are analyzed as elements that are involved in the structuration process of subjects and/or signs. They now gain their own rugged 'facial' expressions within their creative possibilities (ranging from conformist to contesting and contextual alternatives), while they are at the same time exposed to the 'atmospheres' of urban space in an architectural, social, or aesthetic sense. Following Heidegger's expression, 'dwelling' then is not just ontological rooting, but also an aesthetic activity on the surface of social fields, a social poetry. Urban dwellers, who cut their life worlds out of a dense network of society, economic and power relations, or cultural expressions, now pass through aesthetics, emotions, and arts. Their social mosaic then encompasses in the city all classes and social groups, but each piece appears in a different way and/or combination.

In this context, autonomous housing in favelas, bidonvilles and shanty towns is aesthetically (and emotionally) as authentic, even fascinating, in its social positioning, as the styling of elite housing and social and commercial activities by starchitects (e.g. in Brasilia, Las Vegas, Dubai etc.). Also governmental action is interpreted as moulding parts of the city within such a perspective, procuring its own forms of limited performance through facial activities. Consequently, social conflicts arouse not only in disputes over produced space, embedded in capitalist power relations or consumer chances, but also over the choices of how and where to expose meanings and how and when to appeal to specific subjectivities. Hereby, the aspects are configured via atmospheric configurations that mediate emotions and feelings in the everyday world of urban dwellers beyond systemic value networks. In this sense, the city is not just the urban plane of networks and relations, but also the amalgamated psychogeography of a half-thing landscape that is composed of differentiated senses and philosophies of dwelling in a poetic sense. The 'right to the city' (Lefebvre 1996), therefore, is not only the guaranteed access to urban space itself, but also the possibility to define relevant urban dimensions. Thus, the discussion on aesthetic theories paves the way to a geography that targets space as an expression of full life in its broadest sense and forms.

References

Adriani, G., Konnertz, W. and Thomas, K. 1981. *Joseph Beuys. Leben und Werk.* Cologne: DuMont.

Anderson, B. and Harrison, P. (eds) 2010. *Taking-Place. Non-representational Theories and Geography.* Aldershot: Ashgate.

Arendt, H. 1958. *The Human Condition.* Chicago: University of Chicago Press.

Bachelard, G. 1994. *The Poetics of Space.* Boston: Beacon Press.

Bachtin, M. 1979. *Die Ästhetik des Wortes.* Frankfurt am Main: Suhrkamp.

Bærenholdt, J.O., Haldrup, M., Larsen, J. and Urry, J. 2004. *Performing Tourist Places.* Aldershot: Ashgate.

Bakhtin, M. 1990. *Art and Answerability: Early Philosophical Essays.* Austin, TX: University of Texas Press.

Baudrillard, J. 2001. *Simulacra and Simulation.* Ann Arbor: University of Michigan Press.

Böhme, G. 1995. *Atmosphäre. Essays zu einer Neuen Ästhetik.* Frankfurt am Main: Suhrkamp.

Caetano, K.E., Duarte, F. and Ferrara, L.d'A. (eds) 2007. *Curitiba: do Modelo à Modelagem.* São Paulo: Annablume.

Castro Nogueira, L. 2009. 'Bubbles, Globes, Wrappings, and Plektopoi: Minimal Notes to Rethink Metaphysics from the Standpoint of the Social Sciences'. *Environment and Planning D: Society and Space*, 27(1), 87–104.

Debord, G. 1966. 'The Decline and the Fall of Spectacle Society'. *The Situationist International* [Online]. Available at: http://www.cddc.vt.edu/sionline/si/ decline.html [accessed: 1 May 2010].

Deleuze, G. and Guattari, F. 1988 [1980]. *A Thousand Plateaus: Capitalism and Schizophrenia.* Translated from French by B. Massumi. New York: University of Minnesota Press.

Delgado, P. and Deschamps, M.V. 2009. 'Região Metropolitana de Curitiba: mudanças na estrutura sócioespacial no período 1991–2000', in *Dinâmicas Intrametropolitanas e Produção do Espaço na Região Metropoltana de Curitiba*, edited by R. Moura and O.L.C. de Freitas Firkowski. Rio de Janeiro: Observatório as Metrópolis, Curitiba: Letra Capital, 211–32.

de Oliveira, D. 2000. *Curitiba e o Mito da Cidade Modelo.* Curitiba: Universidade Federal do Paraná (UFPR).

de Souza, M.L. 2006. *A Prisão e a Agora. Reflexões em Torno da Democratização do Planejamento e da Gestão das Cidades.* Rio de Janeiro: Bertrand Brasil.

Dias, S.I.S. 2006. *A Arquitetura do Desejo. O Discurso da Nova Identidade Urbana de Curitiba.* Cascavel: Assoeste.

Elden, S. and Mendieta, E. 2009. 'Being-with as Making Worlds: The "Second Coming" of Peter Sloterdijk'. *Environment and Planning D: Society and Space,* 27(1), 1–11.

Fischer-Lichte, E. 2004. *Ästhetik des Performativen.* Frankfurt am Main: Suhrkamp.

Franck, G. 1998. *Ökonomie der Aufmerksamkeit: Ein Entwurf.* München: Carl Hanser Verlag.

Gil, A.C. de Freitas 2003. *Shopping Centers em Curitiba. Produção de novos Espaços de Consumo.* Master thesis in Geography. Curitiba: Federal University of Paraná.

Gottdiener, M. 1995. *Postmodern Semiotics. Material Culture and the Forms of Postmodern Life.* Oxford: Blackwell Publishers.

Guattari, F. 1995 [1992]. *Chaosmosis. An Ethico-Aesthetic Paradigm.* Translated from French by Paul Bains and Julian Pefanis. Indiana University Press.

Haesbaert, R. 2004. *O Mito da Desterritorialização. Do 'Fim dos Territórios' à Multiterritorialidade.* Rio de Janeiro: Bertrand Brasil.

Harvey, D. 1973. *Social Justice and the City.* Baltimore: John Hopkins University Press.

————— 2009. *Social Justice and the City.* 2nd Edition. Athens: University of Georgia Press.

Haug, W.F. 1986 [1971]. *Critique of Commodity Aesthetics.* Minneapolis: University of Minnesota Press.

Heidegger, M. 1993 [1926]. *Sein und Zeit.* Tübingen: Niemeyer.

————— 2001a [1951]. 'Poetically Man Dwells', in *Poetry, Language, Thought,* edited by M. Heidegger. New York: Perennial Classics, 209–28.

————— 2001b [1951]. 'Building, Dwelling, Thinking', in *Building, Language, Thought.* New York: Perennial Classics, 141–60.

————— 2005 [1935]. *Der Ursprung des Kunstwerks*. Stuttgart: Reclam.

Hjelmslev, L. 1971. *Essais Linguistiques*. Paris: Minuit.

Irazábal, C. 2005. *City Making and Urban Governance in the Americas. Curitiba and Portland*. Aldershot: Ashgate.

Kazig, R. 2007. 'Atmosphären – Konzept für einen nicht repräsentationellen Zugang zum Raum', in *Kulturelle Geographien. Zur Beschäftigung mit Raum und Ort nach dem Cultual Turn*, edited by C. Berndt and R. Pütz. Bielefeld: transcript, 167–188.

Lefebvre, H. 1991 [1973]. *The Production of Space*. Oxford: Blackwell Publishers.

————— 1996. 'The Right to the City', in *Writings on Cities*, edited by E. Kofman and E. Lebas. Malden: Blackwell Publishers, 63–181.

————— 2003 [1970]. *The Urban Revolution*. Minneapolis: University of Minnesota Press.

Lorimer, H. 2005. 'Cultural Geography: The Busyness of Being "More-than-Representational"'. *Progress in Human Geography*, 29(1), 83–94.

Maffesoli, M. 1990. *Au Creux des Apparences. Pour une Éthique de l'Esthétique*. Paris: Plon.

————— 1996. *Éloge de la Raison Sensible*. Paris: B. Grasset.

————— 2007. *O Conhecimento Comum. Introdução à Sociologia Compreensível*. Porto Alegre: Sulina.

Marx, K. 1990a [1844]. *Ökonomisch-philosophische Manuskripte aus dem Jahre 1844*. Marx-Engels-Werke, vol. 40. Berlin: Dietz Verlag, 465–588.

————— 1990b [1847]. *Das Elend der Philosophie. Antwort auf Proudhons, Philosophie des Elends*. Marx-Engels-Werke, vol. 4. Berlin: Dietz Verlag, 65–182.

————— 2005 [1867]. *Das Kapital. Kritik der politischen Ökonomie. Erster Band: Der Produktionsprozeß des Kapitals*. Marx-Engels-Werke, vol. 23. Berlin: Dietz Verlag, 3–801.

Marx, K. and Engels, F. 1990 [1847]. *Die Deutsche Ideologie. Kritik der neuesten deutschen Philosophie in ihren Repräsentanten Feuerbach, B. Bauer und Stirner, und des deutschen Sozialismus in seinen verschiedenen Propheten*. Marx-Engels-Werke, vol. 3. Berlin: Dietz Verlag, 11–530.

Menezes, C.L. 1996. *Desenvolvimento Urbano e Meio Ambiente. A Experiência de Curitiba*. Campinas: Papirus.

Morin, M.-E. 2009. 'Cohabitating in the Globalised World: Peter Sloterdijk's Global Foams and Bruno Latour's Cosmopolitics'. *Environment and Planning D: Society and Space*, 27(1), 58–72.

Moura, R. and Firkowski, O.L.C. de Freitas (eds) 2009. *Dinâmicas Intrametropolitanas e Produção do Espaço na Região Metropoltana de Curitiba*. Curitiba: Letra Capital.

Müller, J. 2004. *Elementos Semióticos no Planejamento Urbano. O Caso de Curitiba*. Master thesis in Geography. Curitiba: Federal University of Paraná.

Sahr, W.-D. 2001. 'Between "Metaphysica" and "Spectaculum". The transformations of the religious element in the City of Curitiba, Brazil', in *El Espacio en América Latina: el contrapuento entre lo local y lo global*, edited by B. Lisocka-Jägermann. Warsaw: Centro de Estudios Latinoamericanos, University of Warsaw, 137–45.

———— 2009. 'Portos e sertões – reflexões sobre uma geografia cultural à la brésilenne', in *Espaço e Tempo. Complexidade e Desafios do Pesnar e do Fazer Geográfico*, edited by F. Mendonça, L.C. Löwen Sahr and M. da Silva. Curitiba: Ademadam, 261–88.

Sánchez García, F.E. 1997. *Cidade Espetáculo: Política, Planejamento e City Marketing*. Curitiba: Palavra.

Sánchez, F. 2003. *A Reinvenção das Cidades para um Mercado Mundial*. Chapecó: Argus.

Schmid, H. 2009. *Economy of Fascination: Dubai and Las Vegas as Themed Urban Landscapes*. Urbanization of the Earth, vol. 11, Berlin, Stuttgart: Bornträger.

Schmitz, H. 2003. *Was ist Neue Phänomenologie?* Rostock: Koch Verlag.

———— 2009. *Kurze Einführung in die Neue Phänomenologie*. Freiburg, München: Karl Alber.

Seel, M. 2000. *Ästhetik des Erscheinens*. Frankfurt am Main: Suhrkamp.

———— 2007. *Die Macht des Erscheinens*. Frankfurt am Main: Suhrkamp.

Simmel, G. 1996. *Philosophie des Geldes*. Gesamtausgabe, vol. 6. Frankfurt am Main: Suhrkamp.

Sloterdijk, P. 1986. *Kopernikanische Mobilmachung und ptolemäische Abrüstung*. Frankfurt am Main: Suhrkamp.

———— 1998. *Sphären, vol. 1. Blasen. Mikrosphärologie*. Frankfurt am Main: Suhrkamp.

———— 1999. *Sphären, vol. 2. Globen. Makrosphärologie*. Frankfurt am Main: Suhrkamp.

———— 2004. *Sphären, vol. 3. Schäume. Plurale Sphärologie*. Frankfurt am Main: Suhrkamp.

———— 2005. *Im Weltinnenraum des Kapitals*. Frankfurt am Main: Suhrkamp.

Thrift, N. 1996. *Spatial Formations*. London: Sage Publications.

———— 2007. Non-Representational Theory: Space, Politics, Affect. London: Routledge.

van Tuinen, S. 2006. *Peter Sloterdijk: Ein Profil*. Paderborn: Fink.

Waldenfels, B. 2004. *Phänomenologie der Aufmerksamkeit*. Frankfurt am Main: Suhrkamp.

Welsch, W. 1990. *Ästhetisches Denken*. Stuttgart: Reclam.

Zirkl, F. 2007. *Die Bedeutung der urbanen Ver- und Entsorgung für eine nachhaltige Stadtentwicklung in Brasilien. Das Fallbeispiel Curitiba*. Tübinger Geographische Studien, vol. 148. Tübingen: Selbstverlag des Geographischen Instituts.

PART III
Implementing Fascination:
Case Studies

Chapter 8

From Dreamland to Wasteland: The Discursive Structuring of Cities

Ulrike Gerhard, Ingo H. Warnke

Introduction

Fascination is a basic feature in numerous modern architectural and city planning debates, often correlated to the exaltation of size, elevation, recklessness or genuineness of architectural plans and their realizations. Particularly in neoliberal enthusiasm it is connoted with an antirational impetus in respect to star architectures, for instance in the field of museum construction (Bilbao-Effect) or in the development of shopping malls.

Fascination, originally signifying 'affect by witchcraft or magic' (OED Online 2010), is usually expressed through extraordinary characteristics. However, fascination often exceeds plain enthusiastic admiration, especially when it is connected to the acknowledgement of new achievements and possibilities. Insofar, fascination also has a utopian potential. It reveals possibilities for the other, it may be newer, bigger, higher, etc. Though utopias imply a transgression of rational borders, such as the utopian models of the ideal cities, they are also rooted through reasoning in the field of fascination. Here, fascination and utopia meet in the anticipating consciousness of the human being rendering possibilities for conceptions that are independent of the concreteness of their realization. This transitory thought of utopia is formulated in its complete broadness in Ernst Bloch's *Das Prinzip Hoffnung* [*The Principle of Hope*] (1959). Bloch, in the context of his neo-Marxist philosophy, is extensively concerned with the fascinating utopias in architectural history, not in the sense of 'the town planning of these unswerving functionalists' (Bloch 1959: 861)[1] but as a 'groundplan of a better world' (ibid.: 872).[2]

As such, fascination in urban construction is to no degree limited to architectural events in their materiality but represents a substratum for urban utopias. Specifically, neo-Marxist town criticism highlights the binary opposition between a reality of failed urban development and the fascination for possible alternatives. Bloch even claims that all genuine architectural achievement is

---

1 'die Stadtplanung dieser unentwegten Funktionalisten.'
2 'Grundrisse einer besseren Welt.'

implemented in the anticipation of an appropriate human space (Bloch 1959: 872) and that therefore architecture in itself possesses socio-utopian content.

The socio-political commitment to urban development in the post-war period can be interpreted from such a perspective. After the break-down of the pre-war world in Europe, fascination in a politically different world defined the reflection on a more humane city development throughout the 1960s and 1970s in broad parts of the Western countries. While Jane Jacobs' book on *Death and Life of Great American Cities* (1961) became a bestseller in North America, advocating new principles of urban development and renovation which should be closer to the residents' needs, in Germany the debate became first and foremost linked to the works of social psychologist Alexander Mitscherlich, namely to his essay on *Die Unwirtlichkeit unserer Städte* [*The Inhospitality of our Cities*] (1965). The socio-utopian reflections of Mitscherlich and other urban critics appeared at a time in which the phase of reconstruction of German cities had been substantially concluded, and the cities were beginning to grow rapidly along their fringes, not least because of the growing wealth. Urban planners, architects and above all sociologists, such as Hans Paul Barth (1961), demanded a disengagement from traditional construction designs by means of modern urban development. So, in essence, it was a matter of innovation through renunciation of overcome construction and settlement patterns, or a renewal of the metropolis through urbanization itself, ergo a conscious commitment to planning for a condensed modern city (Barth 1961: 108). In the writings and debates of those times, urban spaces were associated immediately with the fascination by a possible participatory society. Thus, the Basel-based expert in constitutional law, Edgar Salin, opened the Augsburg Städtetag (Association of Cities) in 1960 with the remark that urbanity cannot work as an isolated spatial formation but needs the participation of residents in a political-social space (Salin 1960).

This applies likewise to North American cities. The aforementioned text by Jane Jacobs can be labelled as symptomatic for such a discussion. While metropolises had hitherto experienced a continuing increase in prosperity, social symptoms of decline could frequently be observed. Therefore, the structural crisis could not be explained without recourse to the social context (cf. Soja 2000). Specifically, the critical theory unfolded a certain fascination in this field, and enlightened urban development was promoted as part of general social debates, delivering sufficient topics for conversation on urban utopias (e.g. Harvey 1973).

In this context, urbanity is not a material feature of cities but a disposition of its residents. Consequently, Beate Binder (2006) includes urbanity as a *moving metaphor* in the field of political utopia. In what follows we investigate how such fascinating redrafts of urban societies actually come into existence, which media they are formulated through, and what their impact is. It soon becomes apparent that it is not by accident that the large number of publications on urban development during the 1960s coincides with a general fascination by the idea of a city of the future. First and foremost, such a city appears in the medium of language and here it is negotiated in discourse, which means that different agents, models and ideas are involved. In this spirit, the sociologist Barth states in his reflections on the modern metropolis that architects often expect from sociologists a utopian model of the future society, for which planners would develop their cities (Barth 1961: 10). Sociologists though, according to Barth, cannot deliver such a model. Architects and town planners should work together with them in order to design better cities. Mitscherlich argues in a similar vein when he states: 'Our cities and our residences are products of the imagination as well as the lack of it, the generosity as well as the narrow obstinacy.' (Mitscherlich 1965: 9).[3] Thus, the ideal city model is never a constant: it is shifted and produced by meanings, perspectives as well as socio-economic conditions.

In the intertextual mesh of declaration and dismissal, statements and criticism, the coordinates of urban spaces are negotiated in the discourse itself. The discursive structuring of cities, the assessment as well as reassessment of urban ideas through the use of language is the central theme of our analysis. Initially, this involves words or phrases such as *suburb, monotony* or *utopian model for cities*, figures of argumentation including the concepts *public vs. private*, which all belong to linguistic worlds in which the field of criticism and urban socio-utopia is situated. Thus, the use of words and the application of particular expressions (key words, stigma words etc.) prompts fascination for certain ideas in the first place. In this sense, projective urban spaces emerge in a language-based discourse and not only in the model making of architects and city planners.

But fascination with utopia has also been a business, and consequently the condensed satellite towns of the 1960s and 1970s emerged from socio-utopian debates on new and modern cities. Here, the provocation of fascination could be seen as argumentative support for big capital expenditure programmes. In this respect, the condensed city can also be reflected in the context of an economy of fascination. While today such an economy refers predominantly to the spectacular buildings of star architects, theme parks or complete city layouts such as Las Vegas (cf. Schmid 2006), it rarely counts for *ordinary cities*. Hence, Basten (2009) rightly criticizes the fact that the contemporary public discussion on building culture primarily focuses on representative buildings instead of developing an *aesthetics of everyday culture*. Towns should be reflected considerably more strongly from their fringes, because that is where most people live and experience architecture (cf. also Basten in this volume). Thus, the current debate on an economy of fascination through star architecture is apparently different from urban development debates in the 1960s. At that time, for example, Barth demanded that planning should start in the private sphere (Barth 1961: 113). For him, the honest city developer should not lend himself to illusions since the metropolitan citizen of today does not dream of a 'New Jerusalem', nor does he want the city as a monument to an idea or a community. His single aspiration would be to be able to live freely and decently with his family.

---

3 'Unsere Städte und Wohnungen sind Produkte der Phantasie wie der Phantasielosigkeit, der Großzügigkeit wie des engen Eigensinns.'

Consequently, we assume that the linguistic-communicative substratum of the socio-utopian city has been of great importance to the understanding of success and failure of city development in recent decades. These arguments even have an influence on the present debate on neoliberalism and urban development, always highlighting the positive connotation of utopia in this context. Our interdisciplinary approach, however, focuses on both the positive and negative impacts of utopian construction in urban development, thus pursuing the investigation of the utopian function of discourse itself in an analytical way. In what follows, we first tackle abstract and theoretical questions in this respect, but then pursue an analytical and empirical approach by using the example of Emmertsgrund in Heidelberg, a German model area of condensed living in the 1970s. In this context, we understand our reflections as a contribution to the question as to why the social utopias of the 1960s, which were conceived as fascinating, are so frequently now considered as failed.

Urban Space as a Product of Negotiation, Controversy and Evaluation

If one considers urban spaces in their relation to discursive negotiation and evaluation, one should differentiate between purposes and effects of linguistic action. In our understanding, *purposes* can be interpreted as the intended goals of communicative behaviour, while *effects* are the unintended but effective consequences of communicative action. The distinction between the intended (conscious) function of social action and the unintended (unconscious) consequences has already been drawn by Merton (1957). In the twentieth century, most urban debates were based on the socio-utopian general principles of urban development, namely to consciously improve the living conditions of urban dwellers, generally in the framework of functional solutions for specific planning problems. However, the effects of these debates frequently contradict such intentions, and therefore there is often no substantial evaluation of the cities themselves, but only an evaluation in relation to discursive positions. So, for instance, the general outline of the *garden city* is an intended utopia, but appears obsolete against the background of the *necessity* of a *car-friendly city*, and the intended concept of *condensed living* is opposed to the individual longing for private space and homeownership at the edge of the city, most often depicted as *urban sprawl*.

In this context, urban development is an expression of contradictory discourses. The semantics of urban spaces are not static but marked by shifting contexts conditioned through changing general principles of urban development. As Lyotard (1993) points out, the struggle for the prerogative of interpretation and the allocation of *correct* perceptions is an expression of postmodern knowledge configuration per se and thus such communication can be seen as a struggle in which linguistic games are the means to establish social relations in societies. The disagreement, for instance, between the functionalism of Modern Urbanism, the social utopias of German post-war debates and the North American precursors of postmodern urban theory, personified by the names of Mies van der Rohe, Alexander Mitscherlich and Robert Venturi, represent an intertextual network of statements where social identity can be positioned based on different utopian concepts (cf. Bollmann 2001: 197, Warnke 2006). In such a network of fixed principles and time-bound fascination scenarios, including the promises of solutions for social problems through urban development, urban spaces surely express much more discursive negotiations and evaluations than a simple material environment.

Therefore, the consideration of the semantic struggles (cf. Felder 2006) for city development models is a condition for the dynamic assessments of urban spaces (cf. Gerhard and Warnke 2007). In this sense, urban environments are not simply given objectively but they are culturally produced meanings (cf. already Berger and Luckmann 1966). Helbrecht (2003: 149f.) speaks of a *doppelten Konstruiertheit* [double design] for cultural geographic objects. Here, language is the medium and means of attributing significance in the urban environment. It seems that the actual boom of discourse analysis and poststructuralist approaches responds to these findings, especially within the germanophone debate on the *Neue Kulturgeographie* [New Cultural Geography] (cf. Gebhardt et al. 2004: 305f.). The research results of this debate point to the linguistic construction of normativity, facticity and truth in order to reveal the discursive anchorage of city development, politics and economy. However, Weichardt (2008: 374ff.) rightly observes that the common reference to discourse linguistics following Foucault (cf. Warnke and Spitzmüller 2008, Warnke 2007) or other poststructuralist positions is only selective and eclectic, usually curtailed in large parts and therefore more linked to metatheories than concrete debates. It seems to us that sometimes the debates concerning discourse theory, which decades ago began in other disciplines such as the social sciences and language studies, is a restaged 'exaltation of a habitual otherness' (Dörfler 2005: 67),[4] often being the primary objective for the proclaimed rethinking and epistemological turns in scientific debate.

In the case of urban spaces this means that what is said and thought about cities is not to be separated from what can be visibly experienced in action in the cities. Here also the purposes and effects of negotiation and assessment are an essential element for urban patterning, as planning processes include controversies about urban utopias and emotional preferences. Thus, fascination (and its opposite, rejection) also becomes a fundamental element of urbanity, specifically in four pivotal dimensions: in aesthetics the architecture of buildings induces discursive elements, in psychology it involves emotional relations, for example, through personal preferences for certain architects or life styles, in the lived experience of urban dwellers the discursive dimension is tackled when they establish their life in buildings and streets, and in sociology it is urban practices that define the social relations as part of an unspoken discourse. The shift of the evaluation of planning projects of the 1960s and 1970s in Germany related to the development of both inner cities and satellite towns clearly reveals how – in a kind of paradox assessment

---

4   'Exaltierung eines habituellen Andersseins.'

– the fascination for a modern urban utopia has turned into the perception of the same space as an urban wasteland. This has become the intriguing question for our analysis, which we want to follow up in a case study of the modernist model area of Emmertsgrund in Heidelberg, famous at the time.

Reassessing Urban Space: The Paradoxical Shift from Dreamland to Wasteland

After the 1960s and 1970s, the implemented policies of condensed living led to an inversion of the socio-utopian fascination of planners, to real disappointment, due to the profound devaluations of the newly established urban quarters. Such an evolution can be traced analytically in the example of the Emmertsgrund quarter at the peripheral rim of the old town of Heidelberg in southern Germany. The site was designed and prepared for a large housing complex in a former military area by the end of the 1960s, construction itself started in 1973. To prevent the already known deficiencies of modern urban development projects in other cities, various aspects of community and social development were taken into consideration during the planning process. In this respect, the planning committee was complemented by an interdisciplinary advisory council that opened the discussion to the public. Among the members of this council was the abovementioned social psychologist Alexander Mitscherlich, who had already been actively involved in the debates on urban development on a national level, when he published his pamphlet on the *Inhospitality of our cities* (1965). Now he wanted to contribute in a practical way, especially as he had been a Heidelberg citizen for a long time.

The project of Heidelberg-Emmertsgrund qualifies particularly for an interdisciplinary discourse analysis on urban space, as the district was developed in response to the abovementioned socio-political debates of the 1960s/1970s, which were concerned with a more humane urban development. Its mantra was *urbanity through density*, so that its urban design advocated forms of enclosed, relatively densely populated districts that deliberately contrasted with the rural as well as the beginning suburban landscape (cf. Düwel and Gutschow 2005: 197ff.). Urbanity here was applied explicitly as a design element for modern building. In this sense, it provided a clear boundary to its surrounding area at the fringe of the city. Condensed living was therefore considered a utopian ideal in order to communicate a sense of community to the residents and to provide access to adequate community facilities (ranging from children's playgrounds to shops and important leisure facilities such as city libraries and public swimming pools). Public spaces, where people could meet, were regarded as essential. Additionally, the project followed the Athens Charter, which demanded a rigid separation of functions, a conscious avoidance of clashes of utilization as these were common at that time. In Germany, some of the first implementations of such projects were the large housing complexes of Mannheim-Vogelstang and Berlin-Gropiusstadt in the first half of the 1960s, which were projected for 20,000, and 50,000 residents respectively. However, as soon as the districts were completed, their prestige turned from fascination to devaluation, and substantial desolateness could be observed. Thus, these development projects soon sharply contrasted with the general expectations, and so they became paradoxical in the literal sense: while conceived as a planned utopia, they triggered disappointment and frustration both for the inhabitants and, not least, for the planners themselves. Tracing the discursive changes in the debate on Heidelberg-Emmertsgrund from the beginning, it becomes clear that the utopian functions and reality are nearly always discussed in contrastive patterns. Thus the development of new town districts was discursively introduced to the public at a time when the academic debate concerning urban development in contrast to the lived urban environments of that time was arising. Thus the discussion itself became an essential agent for the success and failure of urban building ideals. In the course of the debate, different agents acquired different powers.

Such a process of assessment and reassessment can easily be recognized in statements on Heidelberg-Emmertsgrund. Thus the local newspaper, the *Heidelberger Tageblatt*, reported in October 1963 on two readings by Alexander Mitscherlich at the Annual Conference of the German Werkbund, Baden-Wuerttemberg. In these the author suggested that the town was to become the *Geburtstort der Freiheit* [birthplace of freedom], a place in which 'the human being does not succumb to the city as he once did to nature, but carries it into his consciousness' (cf. Heidelberger Tageblatt 1963).[5] According to this report, such an attitude had been neglected up to that time. As can be seen, statements of this kind were still characterized by the enthusiastic impetus for innovations in the field of city planning, which also constituted the breeding ground for the subsequent planning activities of Heidelberg-Emmertsgrund. So it is not surprising that the chairman of the public utility housing enterprise succumbed to the fascination by planning: 'We are attempting to create a new district in Emmertsgrund, which redounds to Heidelberg's adornment and can provide a genuine home for our citizens.' (cf. Rhein-Neckar-Zeitung 1967).[6] And Mitscherlich pondered in a handwritten note: 'It is not the plan of a city but a dream of a city I seek to create. The plan will sort itself out' (cf. Alexander Mitscherlich Archive, hand-written note).[7] In the year 1968, Mitscherlich then demanded that a settlement with an urban character should be erected on the Heidelberg-Emmertsgrund terrain, proclaiming: *Keine Schlafstadt* [not a dormitory town!], as mentioned in *Heidelberger Tageblatt* (Haas 1968) and *Keine Gartenstadt* [not a garden city], as Mitscherlich later wrote in a letter addressed to the Federal Minister of Housing and Urban Development, Lauritz Lauritzen (26 May 1970). In such statements,

---

5   '... dass der Mensch dieser Stadt nicht unterliege wie einst der Natur, sondern sie in sein Bewusstsein überführe.'

6   'Wir versuchen, im Emmertsgrund einen neuen Stadtteil zu schaffen, der Heidelberg zur Zierde gereicht und den Bürgern eine wirkliche Heimat werden kann.'

7   'Nicht den Plan einer Stadt will ich zeugen, sondern den Traum einer Stadt, der Plan wird sich schon finden.'

the agonal discourse principle becomes recognizable that defines positions in distinction to stigmatized counter drafts. Thus, the proposed settlement of Emmertsgrund appears in its urban character as an antithesis to the dormitory town and the garden town, the latter having once been proposed as the ideal form of living at the end of the industrialization of the nineteenth century.

In 1968, the decision was made by the advisory council to approve the project of Heidelberg-Emmertsgrund submitted by architects von Branca & Angerer (Munich). The project which mediated between the poles of sociological ideals and economic constraints, was labelled a *Musterplanung* [model design] (Heidelberger Tageblatt 1969) and a 'project of exemplary importance' (Rhein-Neckar-Zeitung 1969).[8]

As early as 1970, the first doubts concerning the success of this seemingly fascinating new suburb arose. Thus the administration of Heidelberg Town Council, after having familiarized itself with similar projects on a trip to the Netherlands and England, expressed concerns that a *good urban development* could simultaneously be a *bad residential construction*. It also suspected in the case of England 'that in a few years those pleasant, new areas such as Thamesmead, a residing city for a total of 60,000 Londoners, will be the new English slums' (Haas 1970).[9] And the fact is that Thamesmead, a district that emerged in Greater London based on the socio-utopian fascinations of the 1960s, later became an issue in the urban debate after it turned out to be a prototype of a problem district (cf. Robbins 1996).

Consequently, the Emmertsgrund project was debated and reconsidered from the beginning of its construction (Heidelberger Tageblatt 1970a). 'A cluster of questions concerning Emmertsgrund' was raised (Rhein-Neckar-Zeitung 1970),[10] the *Platzmangel* [shortage of space] in the area was bemoaned (Heidelberger Tageblatt 1970b) and what had been thought to be the *Stadt der nächsten Zukunft* [city of the near future] (Rhein-Neckar-Zeitung 1972) was now labelled a *Satellitenstadt* [satellite town] which presumably could not 'respond to the targeted expectations' (Stuttgarter Zeitung 1973).[11] The Heidelberger Tageblatt soon declared: 'Nobody is queuing up for apartments there' (1974a).[12] Two weeks later, the same newspaper noticed that there were 'far better opportunities to live in rented accommodation' (Heidelberger Tageblatt 1974b).[13]

In the year 1977, the poor image of the district had been confirmed and the question was posed of how a 'ghetto situation can be prevented' (Heidelberger

---

8   'Projekt von exemplarischer Bedeutung.'
9   'daß in einigen Jahren diese schönen, neuen Stadtteile die neuen englischen Slums sein werden, so in Thamesmead, der Wohnstadt für insgesamt 60.000 Londoner.'
10   'Ein Bündel Fragen zum Emmertsgrund.'
11   'Trabantenstadt Emmertsgrund wird vermutlich die in sie gesetzten Erwartungen nicht erfüllen können.'
12   'Nach Wohnungen steht dort niemand mehr Schlange.'
13   'Schon weitaus besser zur Miete gelebt.'

Tageblatt 1977a).[14] At the same time, the high proportion of migrant residents was also given as a problem, the situation even being compared to Kreuzberg in Berlin (Heidelberger Tageblatt 1979). Ironically, helpless politicians, on visits to the area, demanded a 'loosening up of the housing complex by means of trees' (Rhein-Neckar-Zeitung 1977),[15] thus clearly contradicting the originally praised utopias of a highly condensed city. But eventually, also this kind of comment was criticized, indicating that Emmertsgrund was frequently misjudged (Rhein-Neckar-Zeitung 1980) and that the image of Emmertsgrund itself should become a public issue (Heidelberger Tageblatt 1977b).

Thus, within seventeen years Heidelberg-Emmertsgrund experienced a massive discursive transformation from the utopian *Geburtsstätte der Freiheit* [birthplace of freedom], which the socio-utopian intellectual Mitscherlich had expected (Heidelberger Tageblatt 1963), to the concern about the possible emergence of a *neues Kreuzberg* [new Kreuzberg] (Heidelberger Tageblatt 1979). A net of commentaries, statements, assessments and imputations was drawn over a physically developed surface, which attributed various meanings to it in the context of constantly altering perceptions on urban development. Here, fascination of urban development and mundane dissatisfaction come close together, just as close as the imputations and assessments of urban spaces in the medium of language and the action in the cities themselves.

We assume that such dynamics not only involve the shift of assessments themselves, but also represent a shift between discursive layers, an observation which makes the transformation of fascination into dissatisfaction a structurally describable fact. Thus, the urban socio-utopias of the 1960s, for which Alexander Mitscherlich's ideas are an excellent example, are not simply individual historical statements, but they are dependent on, and reflect, expressive and impressive values of the urban landscape where the pattern of the city is also a discursive space. Here, the theoretical question must be posed as to the structural conditions of such changes.

Structuring the Discursive Debate on Urban Development

It is the process of alternating appreciation and devaluation, fascination and rejection of urban arrangements that is of central interest in our investigation. Heidelberg-Emmertsgrund, like Thamesmead, can be seen as an example of how the urban development model of 'urbanity through density' has caused frustration and even inoperativeness in urban development itself. We propose that the structural revaluation of key concepts is intertwined with new forms of social living, defined in a space of discursive changes in and between intellectual discourses and specific experiences of urban dwellers. Consequently, the discursive enunciations rapidly

---

14 'Einer Ghettosituation vorbeugen.'
15 'Auflockerung der Bebauung durch Bäume.'

disappear behind objections and criticism both in horizontal and vertical discourse dimensions, in a space between ideas and voices.

In our case, we understand the horizontal discourse dimension as the structure of urban planning and assessment, which evolve in historical processes. Here we distinguish three diachronic aspects: (1) *planning* as the conceptual phase, where the development of urban spaces is controversially discussed between agents in the planning process, (2) *implementation* as the phase of active execution of concepts framed by economic and social necessities, and (3) *acceptance* as the phase where the constructed buildings are incorporated into the life-worlds of residents so that differentiated interpretations arise both from the insider's and outsider's perspectives in a wide range of actors' groups.

The vertical discourse dimension in our case refers to the broadening of the social base of the discourse; here we investigate the stratification of personal and institutional agents. In the concert of these different agents, voices can be heard trying to find their audience. We refer to the concept of *Voice* that Blommaert (2005) adopted from Hymes (1996). Both Hymes and Blommaert understand by *Voice* the ability to make one's voice heard in situations where the speakers try to achieve a self-imposed communicative goal. Voice then is defined in general as 'the ways in which people manage to make themselves understood or fail to do so.' (Blommaert 2005: 68). Such a concept can be combined with the premises of action theory in geography and the analysis of geographic agents: under these premises, intentional action is the appropriation and shaping of the world by subjects in a way which geographer Benno Werlen (2007) labels as *performing geography*. In this context, language becomes the expression of subjective rationality as a medium for the production of spaces and regions in action theory, closely bound to the productive and legitimating elements of these places. Werlen's research approach on everyday regionalizations (2007) uncovers three different dimensions (signification-information, domination, and legitimation) and distinguishes especially informative-significant interpretations as essential to the production of space or, in other words, his approach on the production of space grounds in the symbolic, emotional and subjective attribution of meaning, which is predominantly present in speech acts (e.g. Schlottmann 2005). This notion of action provides the possibility for an analytical integration of physical-material circumstances, the state of consciousness and the social world of the agents (cf. Weichhart 2008: 275). Here subjects – including institutionalized communities of subjects in social systems – are agents for the development and formation of urban spaces, and their discursive power emerges due to their actions, the language as well as the attribution of meaning, which finds its expression in the physical-material world, ergo in the specific urban space. In the words of our voice theory: the audibility of voices is regulated in the discursive field of urbanity in various ways, resulting from specific power relations in specific periods of time, so that different levels of expression are activated or deactivated through restrictions that give access to the socially audible sound of a specific urban discourse and its material field.

In the case of the urban space of Heidelberg-Emmertsgrund the discursive structuring and revaluation comes into effect within the two aforementioned discourse dimensions. First, it was the leading intellectual figure of the urban discussion in the 1960s, Alexander Mitscherlich, who created an elevated audibility particularly during the *planning phase* of the Emmertsgrund project. His status as one important proponent for the stratification of the discourse was initially undisputed. The same applied for some media, such as the *Heidelberger Tageblatt*, and political authorities like the Heidelberg Town Council, who spurred on the ideas of urban development in Heidelberg, fascinating equally planners, politicians and potential residents with enthusiasm for the new suburb. In addition to these public debates, analytical material was also provided, such as the minutes of the advisory committee meetings on Emmertsgrund. Controversy arose when the architectural drafts were submitted for the architectural competition. Here, for example, the project of the architect Albert Speer jun. was dismissed in favour of the outline of von Branca & Angerer. Speer's project was believed to cause 'adverse effects on encounters' of the population that would rather comply with residents' desires for 'offering them anonymous living and fast car driving, but abandoning the constraints of the community' (Rhein-Neckar-Zeitung 1968).[16] To strengthen such arguments for an arcadian, but urban community life (for the rural bias in urban geography, see Jacques Lévy in this book) experiences from other cities were also mentioned where progress in urban development was evaluated as having failed. Thus Mitscherlich described a *bürgerkriegsähnliche Zustände* [situation close to civil war] in the Märkische Viertel in Berlin (letter from Mitscherlich, ca 1969/70, addressee unknown), while he projected the Emmertsgrund to be a different and superior *Arcadia*, where fully aired public squares united the residences grouped around the squares. So, the district could become a *Aushängeschild* [flagship] or *Musterstück* [model case] providing new ideas (letter to Mitscherlich, April 1968) that were reinforced with the help of the media and through the publicly effective staging of some figureheads. Such a dreamland contrasted sharply with all the social problems of that time, but provided a pansocietal contribution.

The discursive formation was completely changed, however, through the ensuing *implementation phase*. It immediately became obvious that the Arcadian utopia had to adjust to the economic realities and constraints. Some letters that were exchanged between members of the advisory council and the projects' architects shed particular light on this matter. The chairman of the house building corporation apologized to the urban visionary Mitscherlich in a letter that he had to 'listen to the mundane explanations that referred to the physical limits of housing and urban development' promising that 'even though we have to respond to so many of your good ideas with references to specific complications, a good deal of these ideas will still be found in the constructed reality of Emmertsgrund' (letter from

---

16   'Der Entwurf … ist begegnungsfeindlich … wenn er ihnen anonymes Wohnen und schnelles Autofahren bietet, auf die Zwänge der Gemeinschaft aber verzichtet.'

Geigenbauer to Mitscherlich, August 1969).[17] Mitscherlich, however, bemoaned
that it was 'terrible to witness how all those features of humane habitation in
Emmertsgrund that we have struggled to design are falling victim to the hands of
rationalizing engineers' (letter from Mitscherlich to Geigenbauer, 30 June 1970).[18]
He felt abused as a *abschraubbare Gallionsfigur* [figurehead that can simply
be screwed off and on at will] who had first served as the spearhead for a more
humane urban development, but who was then no longer needed in the process
of its implementation (letter from Mitscherlich to Mayor Zundel, 14 February
1971). The architect and laureate of the project competition himself, von Branca,
was enraged, because it became impossible to realize his multi-storey building
as an architectural project due to 'its obvious simplification by the rationalizing
engineers' (letter from von Branca, 16 July 1970).[19] According to his remarks,
the building constructed was a wrong interpretation of what had been originally
intended. Hence, the realization of these utopias throughout the implementation
phase promptly revealed the limits of urban development: what had been justified
initially as a cross-social objective of the politics of urban development now
came within the scope of the market economy, which thus caused a clear contrast
to the contents of the preceding debates. However, the dispute still remained
predominantly within a circle of experts, while the public coverage by the media
and politicians as yet avoided negative accounts.

It is not until the *acceptance phase* that the whole discursive formation shifted
away from the field of planning agents to the residents themselves. Here it is
interesting that in addition to the internal perspectives of the inhabitants of the
area the perspectives of external citizens are also relevant. It is thus striking that
the evaluations differ considerably between the internal and external group, with
residents almost invariably rating their district more positive than non-residents. In
a survey conducted among 300 households in Emmertsgrund in 2008, 75 per cent
of the residents said that they were content with their district and were pleased to
be living there (cf. Gerhard 2008). Such a high quota can be partially interpreted
as a result of the fact that residents tend to live there for a considerable time,
with roughly three quarters having lived there for longer than five years. These
statements can be linked to the question as to what colour the residents associate
with their district: half of all interviewees spontaneously mentioned the colour

17 'Es tat mir so leid, daß ... Sie sich ... die banalen Erklärungen über die realen
Grenzen des Wohnungs- und Städtebaus anhören mussten. Auch wenn wir bei so vielen
Ihrer guten Ideen mit Hinweisen auf diese oder jene Schwierigkeit antworten müssen, so
wird sich doch noch eine ganze Menge in der baulichen Wirklichkeit des Emmertsgrundes
wieder finden.'

18 'Es war schrecklich zu sehen, wie die Hand der Rationalisierungs-Ingenieure alle
jene Züge von Menschlichkeit des Wohnens im Emmertsgrund zum Opfer gefallen sind, die
wir mit Mühe vorgezeichnet hatten.'

19 'Das Hochhaus ... ist in seiner, offensichtlich durch die Rationalisierungsingenieure
simplifizierenden Vereinfachung dieser Form meines Erachtens unmöglich.'

green, thus pointing to the district's location on the fringe of the town close to a more natural environment. Such a harmonious and even romantic attitude to the life world environment can be interpreted as a positive identification with the district, a fact which is also confirmed by the forms of social involvement of the residents with their living environment (e.g. in the culture association *Kulturkreis Heidelberg-Emmertsgrund* or the district association *Stadtteilverein Emmertsgrund*, in the editorial committee of the district paper *Em-Box-Info*, or in neighbourhood festivals). This stands in clear contrast to the assessments by mass media, and negative evaluations by external residents, both largely influenced by knowledge of the socio-economic data base of the district. Economic prosperity is usually seen as a yardstick for successful urban development. Accordingly, Emmertsgrund appears among those districts that have the highest unemployment rates (17 per cent, while the municipal average is 9.4 per cent), and an extremely high quota of welfare beneficiaries (174 recipients of social assistance among 1,000 residents, in contrast to 59 among 1,000 beneficiaries in the Heidelberg area in general). The district also shows marginal transfer rates for pupils from primary-schools to grammar schools (cf. Stadt Heidelberg 2008). Thus, the evaluation of urbanity in the modern suburb of Heidelberg-Emmertsgrund falls into the discursive divide between experienced reality and promoted ideals, and the different discourse agents and their social position define the semantic struggles for urban spaces. Explicitly, the contrast between an external identity of planners and media, which coincidentally influence the perspectives of outsiders, and an internal identity of the residents is made describable in discourse-semantic basic figures (Busse 1997) via categories of internal/external judgement and social inclusion/exclusion embedded in discourses on urbanity.

These research results clearly demonstrate the complex interaction between the vertical and horizontal dimension of the discourses on urban development. Here, the dynamics of structural assessments are embedded in processes of appreciation or devaluation expressing actual power structures and social stratification. Under this aspect it is remarkable that urban development is incorporated into the utopian ideals in the 1960s and 1970s of individual intellectuals who expressed their ideas with a feeling of fascination for new model cities. Such individualization of reasoning, however, was directed towards an anonymous social world and was based on abstract relations. The physical residents, instead, tended to appear in this situation as a speechless mass and were perceived as external to these discourses, in the worst case even as members of a ghetto. The personified spokesmen of the utopias thus possessed voices that explained the image of urban spaces from the exterior. But the spatial homogenization of their ideals meant a desemiotization of the individual identity of the residents in the life worlds of *condensed living*. Therefore the development programmes of that time promoted *condensed cities* by frequently speaking louder than the residents' voices. And so the supposed consent on community living in a built urban environment of the 1970s was, as such, a semiotic technique of an intellectual elite, where the living subjects had simply vanished. In conclusion, the process of discursive patterning of urban spaces from planning to implementing and living is a socio-semiotic process in which the discursive structural evaluation and acceptance is not only a sign-bound transformation, but a situation in which the discursive dimension of language influences, through theming, the social stratification and even the life worlds of urban dwellers and, as such, urbanity.

Conclusion

In this article we maintain that the assessment of urban spaces as fascinating or disappointing, liveable or unappealing is always embedded in relations to discursive positions. It frequently happens that the intended purposes of planning discourses later deviate completely structurally in their discursive effects during the implementation and acceptance/rejection phase. But our research interest is not so much oriented towards the past. Observing the high degree of discursive formation in the case of condensed living it becomes clear that there is also a great potential for future residents in what are actually still voiceless urban spaces, such as suburbs, social housing areas, ghettos etc., where a discursive revaluation can hamper the further deterioration of mute peripheries, so that sites of silent social misery eventually regain relevance as design spaces for their residents, and even external actors.

This means that in addition to the importance of pansocial debates, socio-political trends and the discursive power of the agents involved, even the architecture of the buildings, the design of public spaces and the events that take place within the built environment (in the sense of *lived buildings*, cf. Lees 2001) also matter. Here, the future task for city planning is not to project a fascinating prestigious object – which sooner than expected becomes a dinosaur of past urban debates – but to build up a sustainable city where urban space can be refined as a living environment by reducing the gap between discursive statements, life worlds and socio-political conditions. Our case study, for example, has shown that public spaces when planned as meeting points or communicational axes, do not invariably fulfil the function assigned to them. They are frequently not able to fill this role, encounter adverse effects (like those mentioned by Mitscherlich), or even deteriorate to becoming a transitory space. This occurs especially, as in the case of Heidelberg-Emmertsgrund, when there is a great discrepancy between the utopian discourses of the planning phase and the later adjustment of the inhabitants in the implemented space. Therefore, fascinating ideas expressed in discourses and model planning only work to a limited extent. If they encounter their limits in the non-discursive world, these limits may be economic, social or political; they frequently become an urban element that is disappointing, and the discourses have to adjust according to social position.

In our opinion, the reconstruction of semiotic evaluation processes between dreamland and wasteland in urban development is scientifically possible by means of an analysis of the discursive construction of urban spaces. Horizontal and vertical dimensions then have to be taken into consideration so that the attribution of meaning can be revealed as a semantic struggle over lived spaces. Such an interdisciplinary approach seems to be of great methodological relevance for further development perspectives of the city.

References

Barth, H.P. 1961. *Die moderne Großstadt. Soziologische Überlegungen zum Städtebau*. Reinbek: Rowohlt.
Basten, L. 2009. 'Überlegungen zur Ästhetik städtischer Alltagsarchitektur'. *Geographische Rundschau*, 61(7–8), 4–9.
Berger, P.L. and Luckmann, T. 1966. *The Social Construction of Reality: a Treatise in the Sociology of Knowledge*. Garden City, NY: Doubleday.
Binder, B. 2006. 'Urbanität als "Moving Metaphor". Aspekte der Stadt-entwicklungsdebatte in den 1960er/1970er Jahren', in *Stadt und Kommunikation in bundesrepublikanischen Umbruchszeiten*, edited by A. von Saldern. Stuttgart: Steiner, 45–64.
Bloch, E. 1959. *Das Prinzip Hoffnung*. Frankfurt am Main: Suhrkamp.
Blommaert, J. 2005. *Discourse. A Critical Introduction*. Cambridge: Cambridge University Press.
Bollmann, U. 2001. *Wandlungen neuzeitlichen Wissens. Historisch-systematische Analysen aus pädagogischer Sicht*. Würzburg: Königshausen & Neumann.
Busse, D. 1997. 'Das Eigene und das Fremde. Annotationen zu Funktion und Wirkung einer diskurssemantischen Grundfigur', in *Die Sprache des Migrationsdiskurses. Das Reden über »Ausländer« in Medien, Politik und Alltag*, edited by K. Böke, M. Jung and M. Wengeler. Opladen: Westdeutscher Verlag, 17–35.
Dörfler, T. 2005. 'Geographie und Dekonstruktion. Zu einem zeitgenössischen Missverständnis'. *Geographische Revue*, 7(1–2), 67–85.
Düwel, J. and Gutschow, N. 2005. *Städtebau in Deutschland im 20. Jahrhundert. Ideen – Projekte – Akteure*. Berlin, Stuttgart: Gebrüder Boerntraeger.
Felder, E. (ed.) 2006. *Semantische Kämpfe. Macht und Sprache in den Wissenschaften*. Berlin, New York: de Gruyter.
Gebhardt, H., Reuber, P. and Wolkersdorfer, G. 2004. 'Konzepte und Konstruktionsweisen regionaler Geographien im Wandel der Zeit'. *Berichte zur deutschen Landeskunde*, 78(3), 293–312.
Gerhard, U. 2008. *Das städtebauliche Leitbild 'Urbanität durch Dichte' auf dem Prüfstand. Eine kulturgeographische Analyse des Heidelberger Stadtteils Emmertsgrund*. Berichte aus dem Arbeitsbereich Anthropogeographie, vol. 9. Heidelberg: Geographisches Institut der Universität Heidelberg.
Gerhard, U. and Warnke, I.H. 2007. 'Stadt und Text. Interdisziplinäre Analyse symbolischer Strukturen einer nordamerikanischen Großstadt'. *Geographische Rundschau*, 59(7–8), 36–42.

Haas, D. 1968. 'Siedlung mit städtischem Charakter. Keine Schlaftstadt! Professor Mitscherlich zur geplanten Emmertsgrund-Siedlung'. *Heidelberger Tageblatt*, 5 May 1968.

———— 1970. 'In England: Auf den Spuren der Emmerstgrund-Siedlung'. *Heidelberger Tageblatt*, 8 July 1970.

Harvey, D. 1973. *Social Justice and the City*. Baltimore: Johns Hopkins University Press.

Heidelberger Tageblatt 1963. 'Die Stadt als Funktion der Selbstdarstellung'. *Heidelberger Tageblatt*, 12/13 October 1963.

———— 1969. 'Heidelberger "Musterplanung" für einen neuen Stadtteil'. *Heidelberger Tageblatt*, 1 March 1969.

———— 1970a. 'Projekt Emmertsgrund neu überdenken. Stellungnahme nach Informationsabend'. *Heidelberger Tageblatt*, 18 July 1970.

———— 1970b. 'Gehört Platzmangel zum Planungssoll?' *Heidelberger Tageblatt*, 19 November 1970.

———— 1974a. 'Nach Wohnungen steht dort niemand mehr Schlange'. *Heidelberger Tageblatt*, 16 August 1974.

———— 1974b. 'Schon weitaus besser zur Miete gelebt'. *Heidelberger Tageblatt*, 31 August 1974.

———— 1977a. 'Einer Ghettosituation vorbeugen'. *Heidelberger Tageblatt*, 10 June 1977.

———— 1977b. 'Image vom Emmertsgrund muß verbessert werden. Kommunalpolitiker erörtern mit Bürgern Stadtteilprobleme'. *Heidelberger Tageblatt*, 18 May 1977.

———— 1979. 'Kein neues Kreuzberg. Neue Heimat zur Mieterstruktur im Emmertsgrund'. *Heidelberger Tageblatt*, 24 November 1979.

Helbrecht, I. 2003. 'Der Wille zur "totalen Gestaltung": Zur Kulturgeographie der Dinge', in *Kulturgeographie. Ansätze und Aktuelle Entwicklungen*, edited by H. Gebhardt, P. Reuber and G. Wolkersdorfer. Heidelberg: Spektrum Akademischer Verlag, 149–170.

Hymes, D. 1996. *Ethnography, Linguistics, Narrative Inequality: Toward an Understanding of Voice*. London: Taylor & Francis.

Jacobs, J. 1961. *The Death and Life of Great American Cities*. New York: Random House.

Lees, L. 2001. 'Towards a Critical Geography of Architecture. The Case of an Ersatz Colosseum'. *Ecumene*, 8(1), 51–86.

Lyotard, J.-F. 1993. *The Postmodern Condition. A Report on Knowledge*. Translated from French by G. Bennington and B. Massumi. Minneapolis: University of Minnesota Press.

Merton, R.K. 1957. *Social Theory and Social Structure*. London: Cohen & West.

Mitscherlich, A. 1965. *Die Unwirtlichkeit unserer Städte*. Frankfurt am Main: Suhrkamp.

OED Online 2010. *Oxford English Dictionary*. [Online]. Available at: http://dictionary.oed.com/ [accessed: 29 March 2010].

Rhein-Neckar-Zeitung 1967. 'Erst den Stall und dann das Haus bauen. Neue Heimat Baden-Württemberg übernahm Emmertsgrund'. *Rhein-Neckar-Zeitung*, 29 December 1967, 3.

————— 1968. 'Obergutachter: Städtebaulich guter Wurf. Martin Speers Emmertsgrund-Entwurf ist eine Alternative zur wissenschaftlichen Wohnkultur der Soziologie – zu groß und zu teuer'. *Rhein-Neckar-Zeitung*, 19 July 1968, 3.

————— 1969. 'Ein Projekt von exemplarischer Bedeutung. Bebauungsvorschlag Emmertsgrund gestern im Gemeinderat'. *Rhein-Neckar-Zeitung*, 31 July 1969.

————— 1970. 'Ein Bündel Fragen zum Emmertsgrund'. *Rhein-Neckar-Zeitung*, 22 July 1970.

————— 1972. 'Impulse für die Stadt der nächsten Zukunft. Baden-Württembergische Stadtplaner im Emmertsgrund'. *Rhein-Neckar-Zeitung*, 18 March 1972.

————— 1977. 'Emmertsgrund schnell ausbauen. Ortsbegehung der CDU empfahl Auflockerung der Bebauung durch Bäume'. *Rhein-Neckar-Zeitung*, 25 June 1977.

————— 1980. 'Emmertsgrund wird oft verkannt'. *Rhein-Neckar-Zeitung*, 16 October 1980.

Robbins, E. 1996. 'Thinking Space/Seeing Space. Thamesmead Revisited'. *Urban Design International*, 1(3), 283–91.

Salin, E. 1960. 'Urbanität', in *Erneuerung unserer Städte*, edited by Deutscher Städtetag. Cologne, Stuttgart: Kohlhammer, 9–34.

Schlottmann, A. 2005. *RaumSprache. Ost-West-Differenzen in der Berichterstattung zur deutschen Einheit. Eine sozialgeographische Theorie.* Sozialgeographische Bibliothek, vol. 4. Stuttgart: Steiner.

Schmid, H. 2006. 'Economy of Fascination: Dubai and Las Vegas as Examples of a Thematic Production of Urban Landscapes'. *Erdkunde*, 60(4), 346–361.

Soja, E. 2000. *Postmetropolis. Critical Studies of Cities and Regions*. Oxford, Malden: Blackwell Publishers.

Stadt Heidelberg (ed.) 2008. *Bericht zur sozialen Lage der Stadt Heidelberg.* Heidelberg: Stadt Heidelberg.

Stuttgarter Zeitung 1973. 'Paradebeispiel mit Schönheitsfehlern. Trabantenstadt Emmertsgrund wird vermutlich die in sie gesetzten Erwartungen nicht erfüllen können'. *Stuttgarter Zeitung*, 21 August 1973.

Warnke, I.H. 2006. 'Die begriffliche Belagerung der Stadt – Semantische Kämpfe um urbane Lebensräume bei Robert Venturi und Alexander Mitscherlich', in *Semantische Kämpfe. Macht und Sprache in den Wissenschaften*, edited by E. Felder. Berlin, New York: de Gruyter, 185–222.

————— (ed.) 2007. *Diskurslinguistik nach Foucault. Theorie und Gegenstände.* Berlin, New York: de Gruyter.

Warnke, I.H. and Spitzmüller, J. 2008. 'Methoden und Methodologie der Diskurslinguistik – Grundlagen und Verfahren einer Sprachwissenschaft jenseits textueller Grenzen', in *Methoden der Diskurslinguistik*, edited by I.H. Warnke and J. Spitzmüller. Berlin, New York: de Gruyter, 3–54.

Weichhart, P. 2008. *Entwicklungslinien der Sozialgeographie. Von Hans Bobek bis Benno Werlen*. Stuttgart: Steiner.

Werlen, B. 2007. *Sozialgeographie alltäglicher Regionalisierungen, vol. 3. Ausgangspunkte und Befunde empirischer Forschung*. Stuttgart: Steiner.

Unpublished Documents from the Alexander-Mitscherlich-Archiv in Frankfurt

Letter to Mitscherlich, addresser unknown April 1968

Letter from Geigenbauer to Mitscherlich, August 1969

Letter from Mitscherlich to Geigenbauer, 30 June 1970

Letter from Mitscherlich to Lauritzen, 26 May 1970

Letter from Mitscherlich, addressee unknown, ca 1969/70

Letter from von Branca, 16 July 1970

Letter from Mitscherlich to Mayor Zundel, 14 February 1971

Chapter 9

Re-designing the Metropolis: Purpose and Perception of the Ruhr District as European Capital of Culture 2010

Achim Prossek

The Ruhr District in Germany, an old region of coal and steel industries that were founded in the nineteenth century, has been made the European Capital of Culture for 2010. However, the promotion of arts and culture can only be seen as a secondary reason for the application for the title. What is more important is the expectation of the emergence of a new image for the area that has taken the centre stage of interest since the big industrial crisis of the 1970s: it is thus hoped that Germany's 'European Capital of Culture 2010' will finally provide a different image of the Ruhr District, different from that of a crisis-ridden environment, associated with decline, based as it was on in the past on the coal and steel industries To promote this image, it is now intended to present the region as a coherent region, in order to fulfil the declared aim to establish a 'Ruhr Metropolis'.

The renaming of the region is part of such a move: no longer is it to be known as the *Ruhrgebiet* [Ruhr District], but rather as the *Metropole Ruhr* [Ruhr Metropolis], thus hopefully marking a transformation from a region of old industries into a post-industrial metropolis. But this re-naming still appears more as an aspiration than a claim or even a realization. We can therefore actually speak of the fact that the Ruhr is being 'staged' as a metropolis. The subject of this chapter is an analysis of this act of staging . I shall venture specifically into the dimension of experience, the aesthetic qualities that give the region its distinct identity, and the emotional foundation that underpins such a 'metropolitanization' strategy.

In this context, the concept of 'atmosphere' is central, encompassing experience, aesthetics, and emotion together. It especially exposes the difficulties and ambivalences associated with the atmospheric reinvention while focusing on the perceptions both of visitors and those living in the region. In determining the image or 'character' of a region, self-image and external perception are two important and interlocking factors intertwined in regard to the appearance and aesthetics of the urban landscape, both in built and immaterial spaces. However, such a crossing-over can become a source of confusion, especially when visitors experience the region. Here, the lack of a clear image can become an obstacle to the growth of culture tourism, to the development of a sense of place and orientation, and to recognizing the aesthetic qualities of the region positively. Consequently, tourism has a special significance for any regional identity and in our specific case this coincides with the creation of a metropolitan image for the region.

The chapter starts by supplying background information on the region and providing an outline of the interests and approaches of key actors, involving the question of how a 'metropolis' can be based on a polystructural environment. Then a discussion on tourism provides an entry into the debate of experience of the region and the impact of mental structuring through attraction and fascination. Then it goes on to the question of what kind of atmospheres are desired and available, and whether they have a metropolitan quality. The paper closes with an examination of the emotional dimension before finally evaluating the act of 'staging' a metropolis as an attempt involving the postmodern planning of fascination.

From 'Ruhr District' to 'Ruhr Metropolis'

The Ruhr District is basically a region of old Fordist industries of coal mines and steel factories. It hosts a population of over 5.2 million in a polycentric urban conurbation without a pronounced hierarchy among its urban centres. Today, it includes eleven cities and four administrative districts with a total of 53 municipalities. Over the last fifty years the decline of the coal and steel industries has led to a loss of some one million jobs. To this day, most of the region's cities remain scarred by above average unemployment rates. Attempts to tie the region to the economic development that is taking place throughout the rest of the state of North Rhine Westphalia have not been altogether successful. After two decades of a policy of pursuing more traditional types of structural programmes, with investments in infrastructure and alternative industries, the late 1980s saw the beginning of a different strategy, implementing a more innovative programme for renewal, the IBA International Building Exhibition Emscher Park. This programme has transformed several disused industrial sites into popular venues for cultural and leisure purposes. However, the core message underlying the IBA programme contained an ecological claim encapsulated in the motto *Change without Growth?*, as was suggested at the 1996 Architecture Biennale in Venice, where the ideas of the IBA were publicly exposed (cf. Wachten 1996). The effect of this attempt was to force a cultural development which culminated in 2002 in the inception of the Ruhrtriennale Festival of Culture, when the locations which reflected industrial heritage were used as a setting for a high-calibre festival (cf. Prossek 2005). Like this festival, the present *Year of a European Capital of Culture* offers hope for greater investment in the infrastructure, an improvement to the region's image, a boost to tourism and greater co-operation within the region.

Connected to these goals is the expectation that such a 'Cultural Year' could strengthen the regional identity of the Ruhr District by endeavouring to foreground the region's metropolitan character. For despite its size, in terms of both population and surface area, the Ruhr District is not perceived either as a single city or as a metropolis. This is regarded as a specific deficit when it comes to attracting both investors and new residents, and is held responsible as one of the reasons for the poor ranking of the region in league tables for cities. All too frequently the media perpetuates such a stereotype of the area, and, despite its gigantic proportions, even highlights the provincial character. As a counter strategy, since 2005 the *Regionalverband Ruhr* [Ruhr Regional Association] has been promoting the *Metropole Ruhr* [Ruhr Metropolis] predominantly for purposes of image promotion and economic development. The aspiration hereby is two-fold: on the one hand, such an image gives expression to a reappraisal of the region in its present shape. Here, greater recognition is given to what has been achieved since the decline of the coal and steel industries, for example, the large and still-growing number of universities, theatres and museums; this increase is held as evidence for the region's development towards a metropolis, even if not much credit is given to their quality and status (cf. Prossek 2008). At the same time, 'metropolis' contains a promise regarding the region's future development: namely a new relationship between the region as a whole and its individual municipalities. Following this interpretation, the *Year of the European Capital of Culture* is viewed as an instrument for integrating regional development and as tantamount to the 'staging' of the region. This is supported by various events and by the creation of a number of 'creative quarters', not merely in the interests of urban and economic development. These are trumpeted as symbols of a more attractive lifestyle, which is seen as an essential element in enabling the Ruhr to hold up in national and international comparison with other cities.

The region also hopes to see improvements at other levels than its promotion strategies. Thus impulses are expected from creative industries that will help to transform some urban neighbourhoods into areas with new homes for the much courted 'creative class'. The 'creative class' is by no means a new phenomenon: as long ago as the late 1980s, a survey of municipal culture policies in Germany referred to culture as a stimulus for the development of distinctive urban neighbourhoods for industries of the future (Wagner 1988: 77). But the current discussion on the creative class frequently ignores a comment made by Sharon Zukin in 1995, which fully applies to the Ruhr District: 'So cultural strategies are often a worst-case scenario of economic development. When a city has few cards to play, cultural strategies respond to the quality-of-life argument that encourages flight to newer regions. They represent a weapon against the decentralization of jobs from established industrial concentrations' (Zukin 1995: 273f.). This is consistent with the programmatic intention of a *Year of a European Capital of Culture*: Originally, in 1985, its policies were designed as a means for promoting art and the arts. But since Glasgow (1990) at the latest art and culture have been viewed within the Capital of Culture framework as an instrument for urban development and city marketing – as an opportunity for a city to 'stage' itself and to attract attention. In this light, the Ruhr District is a 'delayed region'. What is new is the attempt to be perceived as a new European metropolis.

The Difficult Organization of a 'Capital of Culture' in a Polycentric Urban Region

There is no single city being promoted for Ruhr 2010 – it is the entire region. And it is the first time that such an explicit polycentric regionalization has been involved in the context of European 'Capitals of Culture'. Though Lille (2004) and Luxembourg (2007) also, quite deliberately, produced programmes that extended beyond their city boundaries introducing some elements of regionalization, it was Essen, that applied as a single city for the status of a Capital of Culture on behalf of all the municipalities of the surrounding region, presenting itself only as the geographical centre. This has been made explicit in the slogan *Essen for the Ruhr*, in which the equal claims of all 53 municipalities are entailed by the preposition *for*. The successful application is seen as marking a historic and unique success for the region, as a *Glanzlicht regionaler Kooperation* [highlight of regional co-operation] (Betz 2008: 191). The shorthand label *Ruhr.2010* was soon agreed on. Whereas media reports initially referred only to Essen as the applicant, by the end of 2009 it had become common to assign the status of a 'Capital of Culture' to the entire Ruhr District – in contravention of the EU Commission's official designation, which continues to refer only to 'Essen', as the statute does not allow for applications from regions, but only from individual cities (European Commission 2010). Thus one of the key objectives of those responsible for the application has been achieved in advance: the region appears as an entity of its own. As a consequence, events have been planned for practically all towns and cities of the region. The arena is, therefore, the entire urban landscape of these cities in their joint existence. Thus, the 'territory' of the Ruhr District arises in its entirety as representing a frame for a coherent region. Even though some individual locations or municipalities within the region may receive more attention than others, which is most particularly the case for Essen (as standard bearer for the region), the narrative of transformation driven by culture includes all the region's towns and cities. This seems to be entirely appropriate, since they have all long suffered from the still prevailing image of a Ruhr District in Crisis.

The responsibility for the organization and marketing of Ruhr Metropolis lies with the *Ruhr.2010 GmbH* [Ruhr.2010 company]. This company has four shareholders: the City of Essen (with €6m), the Ruhr Regional Association (RVR, €12m), the State of North Rhine-Westphalia (€12m), and the *Initiativkreis Ruhr* [Ruhr Initiative Circle] (€8.5m). While the City of Essen and the State of North Rhine Westphalia are mainly institutional partners, the Ruhr Initiative Circle is an alliance of some 60 nationally and internationally operating companies which have dedicated their activities to supporting the economic restructuring in the Ruhr District during the last 20 years. In turn, the RVR has a majority stake in the *Ruhr Tourismus GmbH* [Ruhr tourism company], and thus is Ruhr.2010's most important single partner. It is this RVR that also commissioned the tourism strategy to accompany Ruhr.2010 and acquired for the region the status of an official partner at the 2009 ITB travel trade fair in Berlin. The total budget for Ruhr.2010 amounts to €65.5m, which puts it roughly on a par with Linz 2009. However, sponsoring has raised significantly less money than originally envisaged. Nonetheless, the programme still contains some 300 projects.

According to its 'Mission Statement for Ruhr.2010', the RVR by working jointly with its partners 'supports the development of sustainable effective structures for the Ruhr cultural metropolitan area. Ruhr.2010 has brought together the regional players from culture, politics and business into a creative alliance. They are all following the ambitious aim to transform the Ruhr metropolitan area into an important player in the future of Europe, and give the area a new and unique brand mark on the map of Europe' (Ruhr.2010 Gesellschaft 2009a). There is no indication as to whether the *leitmotif* of a common metropolitan identity really enjoys the support of the municipalities. However, such a fact is not decisive for the strategy of implementing an image of a Capital of Culture, as such a project focuses only on the attention, much in the way that a magnifying glass concentrates light on one spot, and not on the effective support. Therefore it is not necessary to distinguish between individual towns and cities. It is the way the region itself is represented and staged which is at stake, and not the lived identity. However, mainly smaller and more peripheral municipalities see a danger of being dominated by Essen since they neither boast cultural amenities of comparable standing and prominence, nor are they in a position to host events on a comparable scale. But their fears are also based less on concerns about not profiting to the same degree as the core cities from the envisaged streams of tourists (they are well aware that their appeal lies elsewhere), but they are more concerned that they will not be acknowledged as equal partners. Against this background, 53 municipalities means 53 logics of local policy and forums of discourse.

In consequence, many local councils and associations first understood the term 'metropolis' as suggesting a loss of autonomy, the polycentric structure thus calling for a political strategy which is consensus-oriented. However, for a large number of actors and interest groups, this will mean that a compromise is unachievable because of the great inequality of projects in the different towns and cities of the Ruhr District. In fact, quite different motivations and objectives are in play, but in the case of large-scale projects like Capitals of Culture (or World Exhibitions and major urban development projects) it can almost never be asssumed that the actors involved are pursuing the same interests. To the contrary: according to Klaus Selle, it is inappropriate to speak of an *Gesamtrationalität* [overarching rationality] (Selle 2005: 258) in such a situation. In the case of Ruhr.2010, it is mainly the local authorities that constitute the actors with strong, specific interests, while the private sector is only involved to a very limited extent. Therefore, it is interesting to see that, in spite of the unifying image of Ruhr.2010 and the extremely strong impact of administrative institutions, a lively sense of competition can be observed among all the towns and cities, creating an intricate ambivalence between confirmative imaging and individual action.

The example of the conversion of a former brewery into a centre for culture and creativity in central Dortmund highlights such a situation. In Dortmund, local politicians drew on Ruhr.2010 to help them assert their ideas for the project. All that was originally planned was to relocate an museum of art within the city centre; the council's attitude towards Ruhr.2010 was sceptical. However, since financial assistance from the state government was made contingent on this local regeneration project being embedded within the Ruhr.2010 programme, and since it was possible to make a case for broadening both the substantive and the spatial framework for the local project, the project was duly adopted and gained political backing from the city parliament. The project now features prominently in the Ruhr.2010 programme and thus Dortmund has managed successfully to assert its own specific local interests while at the same time positioning itself as a major city within the Ruhr District.

Another example of the dialectics between local action and regional image is provided by the Ruhr.2010 company itself. Here, the Ruhr Regional Association (RVR) is the majority shareholder and the attempt to reposition the Ruhr District as the Ruhr Metropolis has its roots in the organization's own policy for strengthening the region as the primary field of action. The interest of the RVR in developing a strong Ruhr District might be because a strong Ruhr district promotes the association's own legitimization and safeguards its very existence. Consequently, the more natural it becomes for people to think in terms of the region rather than of individual towns and cities, the more important is the role of a regional authority. This is something which has by no means been uncontroversial in political circles over the last decade. But today, the discourse on a 'metropolis' creates a more urgent need for the logic of legitimization, since it further enhances the status of the region. So, for the RVR, the *Year of the Capital of Culture* represents an additional instrument or medium for 'staging' the metropolis among the population.

Looking at the Ruhr District in its totality, it can be assumed that individual interests exist alongside what might be seen as an overarching rationality of transmunicipal interests. The gain in image is something very important for all local and regional authorities, especially as these currently still suffer from the historic heritage of a negative image of the region. Positive reports are therefore seen as a success, as are increases in the number of visitors to the region. What will be critical, however, is whether individual actors evaluate the achieved co-operation based on a common image as individual gains proportionate to the resources expended. The positive response to this evaluation would be both a binding agent and a stimulus for future developments, and only then will a real chance for regional co-operation exist in the future.

It must be mentioned that the labelling of the region as Ruhr metropolis goes back to a complex legacy, the so-called *Ruhrstadt-Debatte* [Ruhr City debate] (cf. Schmitt 2007: 149ff.). Around 2000, this discussion was polarized primarily around the issue of finding a more effective form of regional administration suited to contemporary circumstances; this restrictive perspective associated with the term 'Ruhr City' triggered a host of negative associations and emotions. However, since the name Ruhr Metropolis even more elevates the concept of city, these negative associations might be reinforced. On the other hand, those people who are propagating the use of the word 'metropolis' work in the fields of culture and economic development, and their proposals appeal to a less concrete and incisive image. This may explain why the term 'metropolis' has not met the same vehement rejection as the term *Ruhrstadt* did between 2000 and 2005.

The soft designing of a regional image can, therefore, induce successful co-operation and strengthen regional cohesion and by these means create the staging of a 'metropolis' that changes both the region's self-image and the regional identity of its inhabitants. A major contributor to the process of creating a regional identity is, in this regard, the tourism industry, as can be seen in the following section.

The Role of Experience for Geographical Imagining: the Influence of the Tourist Gaze on Regional Identity

For tourism, the act of seeing is central, as it is only through seeing that areas of tourism are constructed (Urry 1992, 2002, Pagenstecher 2003, Groys 1997). With regard to the objects of tourism, here 'use' gives way to 'sight' – it is visual consumption. In this context, the 'collection of signs' (Urry 2002: 3) is a prime feature of tourist activity, a semiotic act which would be impossible without preparation through media, and without pictures and images available. Such a process can also be observed in other socio-cultural types of branding. A tourist gaze can therefore be described in our case as a view of the region developed and experienced by those who visit the Ruhr District. Their expectations are primarily bound to the tourists' experience of what was once an industrial region. But the image also includes a second and by no means less important dimension: the way in which the local population re-experiences the distinctive qualities and charm of their own region. Their re-appropriation also takes place through a tourist perspective, as it implies a fresh look at what were ostensibly everyday scenes. The common denominator for both groups, local residents and visitors, is the image they have of a town or city, or of the region, and how they work with it. The common ground for this blurring of the demarcation between 'local resident' and 'tourist' is that both groups share the same typical and characteristic images. This is particularly clear in the case of those towns and cities which are well represented in the media and which attract large numbers of visitors. In the case of Cologne (outside the Ruhr District, and more famous), for example, it is the ensemble created by the Rhine, its bridges, the cathedral and buildings in the historic centre that forms the typical images of films and postcards. In these items the city's image is crystallized and becomes emblematic. For the towns and cities of the Ruhr District, this potential was never made evident, and so the performance of a regional cultural event is still a new phenomenon for both tourists and inhabitants. Again, a tourism perspective requires typical and characteristic sights, and in the Ruhr area these can be found even today in the feature symbols of the industrial heritage. These provide a unique focus for identification of and with the region. By contrast, 'normal' postmodern images of city life would not compete sufficiently with other major cities and metropolises. All that remains for the Ruhr Metropolis is, thus, to emphasize its distinctiveness as being more than a city, namely a specific place of living and scenery.

This has specific consequences for the inhabitants of the region and their attitude towards tourism. Tourism generally entails encroachment of the alien into the familiar, creating a flurrying fascination of attractiveness. Thus, tourism negotiates the boundaries between the inner and the outer space, the elements of self ('typical' and 'characteristic') and of the other. Adopting a tourist's perspective involves perceiving oneself as a potential object, seeing oneself through other people's eyes. And the implications of this may be existential: 'In doing this, we are asking what our own value for tourism is, or precisely what constitutes our own cultural identity' (Groys 1997: 105).[1] By learning to see oneself as an object, one becomes a subject. Thus, in order to be able to act successfully as a supplier, i.e. to have confidence in regarding one's home region as a destination, it is essential to learn how to adopt the tourist's perspective in one's own life world. This means learning to look at one's own familiar surroundings, town and countryside, with the eyes of a stranger. Only then is it possible to experience oneself and one's own region anew. According to Groys, this means travelling frequently, gathering experiences abroad, taking note of the constant renegotiation and movement which takes place in this market, and thus gaining an impression of what lends itself best to what he terms *Selbstmonumentalisierung* [self-monumentalization] and *monumentelle kulturelle Identität* [monumental cultural identity] (ibid.: 107). For Groys cultural identity here amounts to more than an image or a brand: it is the perception of the region as a composite image, a common ground, which predominates among the people of a region. This can only grow through encounters with the outside world.

In remodelling the former Coal Washing Plant of the *Zeche Zollverein* [Zollverein Coal Mine] to create the *Ruhr Museum* (fig. 9.1), Rem Koolhaas has achieved in an exemplary fashion what John Urry (2002: 111) describes as 'designing for the gaze': here the building is designed with a view to its tourism impact, and therefore the tourist gaze is drawn first and foremost to spectacular architecture. The old and the new meet in a particularly striking manner to create a new icon of the Ruhr which immediately becomes popular for its symbolic value. Thus, the *Zeche Zollverein* graces the covers of many of the new travel guides which have been published in connection with Ruhr.2010. Most of the cultural centres and museum buildings which have opened their doors in recent years give recognition to this mechanism in the same way. What is known as the 'Guggenheim effect' comes into play here. Rem Koolhaas is even convinced that such architectural icons actually generate their own audience. They satisfy the need of spectators *nach gebauten Bildern* [for built images] (Koolhaas 2006: 107): here, the image becomes space, and from space comes the image. In the

---

1   'Wir stellen damit die Frage nach unserem touristischen Wert oder nach unserer eigenen kulturellen Identität.'

Figure 9.1   Ruhr Museum at the former Zollverein Coal Mine

case of Ruhr.2010, even events are evaluated in terms of their image potential. There are already concrete plans for a 60-km picnic that is planned to take place on the motorway that runs through the centre of the region ('Still Life A40'), and which it is planned to film from the air – by Ruhr.2010 managing director Fritz Pleitgen, a well-known journalist and former director of the regional public broadcasting station Westdeutscher Rundfunk. This mass event, conceived for media exploitation, is more important as an image than most of the other contents of Ruhr.2010.

For tourists, orientation is always an important issue. They usually try to find their way as easily and quickly as possible. Town halls, museum quarters, shopping streets and market squares often serve as central points of contact – and the route map for public transport provides the grid for mobility. Many (European) cities function along these or similar lines. In this context, the deficits of the polycentric Ruhr region immediately become apparent: the region contains everything and in great number, but it boasts very few outstanding attractions. Thus, the destinations for tourists are spread over the entire region and are consequently far apart; this requires visitors to expend a good deal of time and effort in organizing their travel. For this reason the Ruhr.2010 company was keen to tackle the issue of perception of the region and develop an orientation from the outset. One concern addressed in this new tourism strategy is reducing the region's complexity in order to render it more comprehensible (Ruhrgebiet Tourismus Gesellschaft 2007), thus avoiding problems of orientation.

To this end, the Ruhr.2010 organizers have, together with *Ruhr Tourismus*, opted for two complexity-reducing strategies. The first involves dividing the region into five so-called hubs, each with its own central visitors' centre as a point of contact and departure. The Ruhr's 53 local authorities are thus reorganized through the image structure by five metropolitan gateways. Secondly, the locations for passage projects follow the lines of rivers and public-transport routes, thus facilitating linear orientation. This concept is described in Book One of the 2010 programme as a way out of confusion. Under the motto of *Five for 53: The metropolitan gateways*, it promises that '"Essen for the Ruhr" will be giving this urban cultural network a modern spatial matrix equally comprehensible for both local inhabitants and visitors' (Ruhr.2010 Gesellschaft 2008: 20). Consequently, the five hubs do not have hard and fast boundaries. They cut across towns and cities, counties, and the regions as demarcated for purposes of state-level regional structure policy and within which co-operative, intermunicipal policy-making has been a reality since the late 1980s (fig. 9.2). For local residents, this new spatial matrix, which is intended to make the region more accessible, runs counter to the established structures. To date there are no signs of any improvements of the spatial organization. Instead, in the case of those municipalities whose hinterlands have been dissolved (for example at the administrative district level), there is strong dissatisfaction and a lack of understanding regarding the manner in which this reorganization has taken place. Interestingly, two months before the implementation of Ruhr.2010, the concept of the 'modern spatial matrix' was

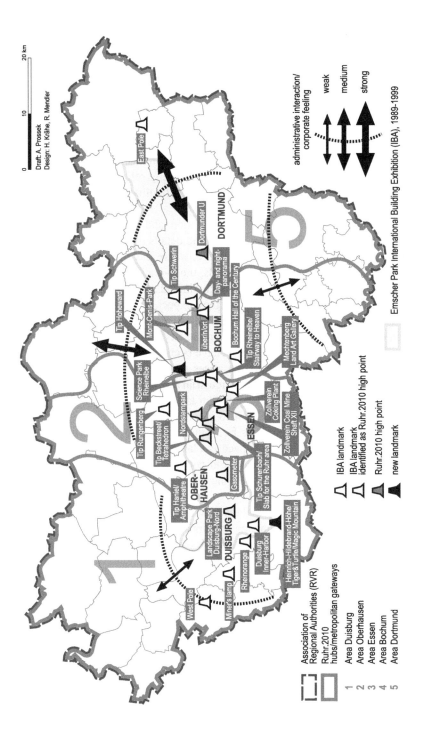

Figure 9.2   Ruhr area: regional structures, landmarks, and spatial identification

Draft: A. Prossek
Design: H. Krähe, R. Mendler

0   10   20 km

Association of
Regional Authorities (RVR)

Ruhr.2010
hubs/metropolitan gateways

1   Area Duisburg
2   Area Oberhausen
3   Area Essen
4   Area Bochum
5   Area Dortmund

△   IBA landmark

△   IBA landmark
      identified as Ruhr.2010 high point

▲   Ruhr.2010 high point

▲   new landmark

      Emscher Park International Building Exhibition (IBA), 1989–1999

administrative interaction/
corporate feeling

weak

medium

strong

dropped: 'The geographical arrangement of these five hubs facilitates orientation and enables all visitors to find what they are looking for within their individual timescales' (Ruhr.2010 Gesellschaft 2009b: 195). The matrix now serves merely as an aid for orientation which divides space according to the amount of time available to visitors, referring to the distance to be travelled between projects, events and sights. The hubs are then no more than a suggestion for making the region more manageable. 'A town has districts. A metropolis has areas' – to quote the programme (Ruhr 2010 Gesellschaft 2008: 17). The aspiration has become both more concrete and more modest: no longer is any reference made to local residents; it is now all about visitors, who are less familiar with the region and need orientation.

What has happened here is the fragmentation and reduction of the total territory into selected zones, and even to specific details depending solely on the visitors' needs. Just as in the case of an image, where the advertising campaign creates a focus by singling out specific details, the overall territory then becomes a secondary goal and is defined only through singled out aspects, in an attempt to guide visitors to make the experiences they expect. For the tourist, this is in fact essential, not least because many of the new leisure and cultural venues are located on old industrial sites well away from the city centres: old dockland areas have been revitalized, new shopping malls have been opened on formerly derelict land. However, a sense of disorientation arises as the new orientation does not follow the old geographical logics of the industrial areas. Even for the visitors, the cities and regions that have lost their function are now no more than 'shapeless conglomerations in which it is difficult to identify a structure and hierarchy of functions' (de Michelis 2006: 16f.). As far as individual experience is concerned, the post-industrial re-organization of space as a landscape for culture and personal experience necessarily entails an increase in complexity, mixing different levels of experience, and hence orientation becomes even more difficult. The Ruhr.2010 organizers have shown an awareness of this problem in developing a concept which addresses the problem of spatial disorientation, but has not yet been able to resolve it. However, there are promising attempts.

One typical way of reassembling an image is by gaining an overview of an area where visiting a location provides a panoramic view. Looking down, the observer creates distance and stands aloof, forming an experience of space through the actual perception of its expansiveness. Such an attitude has tradition (as can be seen with medieval castles, church towers, and skyscraper restaurants): exposed locations that provide panoramic views of a city have always been popular for tourists (Hauser 1990: 107f., Koschorke 1990: 138ff.). As a non-tourist area the Ruhr has rarely provided such opportunity for a panorama. Only in the 1990s, within the context of the Emscher Park International Building Exhibition, was a systematic attempt made to create central points for gaining an overview. The declared objective was to develop people's sense of orientation in a polycentric urban landscape. One initial effect was that, for the very first time, the surrounding landscape was recognized as being cohesive, as forming a single entity. Surprisingly large areas of greenery, coupled with the absence of visible interruptions in the built development, provided a complete picture of the Ruhr District as a region, rather than views of an individual city. The top of the Coal Wash Building of the *Zeche Zollverein* is an impressive case in point, as from here both Essen and Gelsenkirchen (the latter with its famous football stadium of Schalke 04) and even more cities can be seen simultaneously. Such a concentration point has the effect of making the region shrink in perception: its intangible and diffuse proportions take on a visible shape. Thus tourists in their minds develop a general image of the Ruhr that has already fused its towns and cities into a region, but also the inhabitants gain a new insight into their environment from new perspectives.

It can be observed that the longing of many local people to experience a panoramic view of their region is very strong. Slag heaps and other landmarks are extremely popular places to visit and to take visitors to. This can be explained by the altered perspective of the region, which has resulted from the recent transformation the regional economy has undergone. It is only since the end of the dominance of the region's old industries that local residents have been able to appropriate aesthetically the legacy of the sites and urban space they once occupied. Furthermore, a growing interest in regional history has turned many people into explorers of their local surroundings and into tourists on their own doorstep. Awareness of, and identification with, the region is thus triggered in particular by such tourist perspective and panoramic views. Here, we find how successful the selection of panoramic viewpoints is, affirming what landscape planners have claimed: 'Without any one of these locations, nothing today would be the way it is' (Projekt Ruhr Gesellschaft 2005: 191).[2]

Thus, viewing platforms – known as 'High Points' – have a key role to play within the Ruhr.2010 programme. Obviously, the organizers are aware of this. And they do not attempt to put a romantic gloss on the view, but tie it to hardship and suffered reality: 'Here you can see the way the horizon has been shaped and changed by people and machinery, the scars left on the panorama by the coal and steel industries, the features of the ruthlessly exploited natural world, and the attempts to make up for the interventions and heal the scars by planning and re-greening the landscape with the addition of works of art' (Ruhr.2010 Gesellschaft 2008: 56). However, this is not necessarily a metropolitan narrative, especially not of a European Metropolis. What people look at is not a sea of streets and houses, but rather an urbanized industrial landscape (the Ruhr District old style) intertwined by expanses of greenery. This appears more a milder version of Los Angeles (but without the prominent downtown area, and without the variety of the city's distinctive boroughs). Thus, the intense experience provided by these individual 'high points' is artistically enhanced by the perception of an old industrial landscape. This, truly, is very unique, but it is hardly metropolitan.

---

2 'Ohne einen dieser Orte wäre heute nichts, wie es ist.'

Structural Difficulties for Creating Metropolitan Atmospheres

While the aforementioned creation of an image mainly refers to real structures and their perception, the question of metropolis also involves atmosphere. Here, aesthetics are touched on. They not only refer to the question of arts, style and design, but also contain an ecological, environmental dimension. Gernot Böhme (1995) introduced for this the term *neue Ästhetik* [new aesthetics]. In this case, aesthetics are applied to everyday matters, where sensory experience takes precedence over judgement, preferring diffuse corporal-emotional sensations to semiotic mental design. This new focus of aesthetics is placed on 'the relationship between the quality of the surroundings and human well-being' (Böhme 1995: 22f.).[3] It is interesting that Böhme's new aesthetics attaches equal significance to both sides, production and reception: on the producers' side, Böhme presents a general theory of aesthetic work, which reflects the construction of atmospheres; on the recipients' side, he proposes a comprehensive theory of perception (Böhme 1995: 25).

According to Böhme, there are two rules which apply to the way people talk about urban atmosphere: 'Firstly, they mean the atmosphere either from or for the perspective of the outsider. Secondly, this is an attempt to identify what is specific and characteristic of a particular city.' (Böhme 1998: 55).[4] Consequently, the insider's perspective is based on everyday experiences which we do not necessarily think about, while the outsider's perspective refers to the production of such atmospheres. What calls our attention, however, is the unconscious mingling of the dimensions in this process; even if atmospheres are created in our immediate surroundings, they are not necessarily registered mentally, they are simply felt.

Thus an aesthetical perspective broadens the experience of a cityscape beyond the visual dimension and intellectual reflection. Through the immersion into an atmosphere the area is felt and its components are identified and perceived diffusely (Böhme 1995: 15). This concept of aesthetics is capable of capturing the tension between the ambitions of the Ruhr.2010 organizers, who are trying to unite the experiences of visitors and local residents in space by material, but diverse, expressions. An aesthetical approach is thus induced by the strong emotional claim expressed in the name Ruhr Metropolis, which must be realized through *ästhetische Arbeit* [aesthetic work], as Böhme coins it. This means 'endowing things, surroundings and even people with attributes which allow them to communicate something' (ibid.: 35).[5] While the region has little in common with the patterns of

---

3   'Die dabei entstehende neue Ästhetik hat es mit der Beziehung von Umgebungsqualitäten und menschlichem Befinden zu tun.'

4   'Erstens wird von der Atmosphäre in der Regel aus der oder für die Perspektive des Fremden gesprochen. Zweitens versucht man damit, etwas für eine Stadt Charakteristisches zu benennen.'

5   'Dingen, Umgebungen oder auch dem Menschen selbst solche Eigenschaften zu geben, die von ihnen etwas ausgehen lassen.'

a typical European metropolis and its urbanity, the atmosphere people pick up here is largely unspecific. Though a visit to the Ruhr Metropolis falls under the heading of city tourism, the Ruhr District, although Europe's third-largest metropolis after London and Paris, certainly evokes expectations that one does not associate with other European metropolises, or even with a global city. This is particularly the case when advertising presents satellite images of Europe at night with just London, Paris and the Ruhr standing out prominently (for example in Ruhr.2010 Gesellschaft 2008: 54). Such an 'impression' is reinforced by the use of statistics, demonstrating that the region has a large number of museums, theatres, concert halls and universities. However, in terms of the built structure there are only very few areas of the region which display the levels of building density associated with European metropolises. Elsewhere, the dominant impression of the region is much more suburban. Unlike other major industrial cities which emerged in the nineteenth century, such as Berlin, the Ruhr District has never been marked by concentration, but rather by dispersed development and, to this date, these types of settlement continue to typify most of the 'metropolitan' region. This may well provide the basis for a relatively high quality of life, but it is not consistent with the claim contained in the label Ruhr Metropolis. Thus, it is only in a few isolated urban areas that there is any perceptible 'urban flair'.

Consequently, the Ruhr District appears to visitors as a 'dispersed and low contrast periphery, ... a lining up of in-between areas, where all the cities resemble each other, one does not find real borders between city and country' (Salij 2002: 246). This is primarily an outcome of the well-known urban development process of the region that has resulted from a territorially bound industrialization (coal mining and infrastructural lines), and its consequences remain visible to this day: 'a lack of structure and pattern bordering on entropy' has been produced, as Lootsma (2006: 195) put it. Such a state of entropy is seen as a consequence of dynamic changes which have dissolved the traditional structures, spatially as well as socially, of the old industrial society, but where the dispersed (suburban) atmosphere of today that has arisen geographically from this situation, has persisted, with the result that the effects of the actual transformation on the Ruhr society have obviously not yet resulted in the emergence of a new and more 'classical' metropolitan structure.

It is therefore more than important to shape the atmosphere of the metropolis through the attitudes and the behaviour of its inhabitants, something which could originate from emotional tensions between local positioning and visitors expectations. Today, the impression which visitors tend to pick up from local residents is still that the latter classify the region as a 'province' rather than as a 'metropolis': 'I have the feeling that in the Ruhr conurbation there are few places where you sense both density and society' (Ferguson 2006: 44). For tourism, on which hopes are pinned for the future, and for the metropolitan identity, this represents a decisive atmospheric obstacle, at least as far as the objective of international competitiveness is concerned: the absence of a 'globally acknowledged urbanity ... is becoming a key problem for the vision of a Ruhr city which can be compared to leading global metropolises' (Rehfeld 2006: 118). Thus, it becomes apparent how much the visual and atmospheric experience and the experiential appraisal of the area is still conditioned by its inherited structure. For this reason, the Ruhr District is hardly comparable in terms of experience with Paris, London or other major cities. On the contrary, the ambivalence between its metropolitan infrastructure and the predominantly suburban atmosphere has created a landscape of its own, and thus the portrayal of the region as the Ruhr Metropolis clearly refers to a different kind of metropolis, one that it is 'still in the making' (cf. Ruhr.2010 Gesellschaft 2008: 12).

As a consequence, the Ruhr.2010 organizers have attempted to qualify the region by associations evoked from the term 'metropolis' in a different way. Firstly, they are trying to position the region outside of conventional categories, and, secondly, they assume a success that cannot yet be quantified. This, in fact, depends on a new understanding of an atmosphere of difference. The organizers have accepted so far that there is a fundamental difference between the region and other 'typical' metropolises, claiming for their programme affirmative attitudes towards its contradictions. This breathes life into the concept: 'Its attractive and its ugly side, high culture and the "Love parade", local pub and Michelin star restaurant, allotment and landscape park, concert hall and football stadium are never far from each other, and so form a dynamic dialogue of differences' (Ruhr.2010 Gesellschaft 2009b: 4). Thus, the planners refer to what Jacques Lévy has called serendipity in a metropolitan atmosphere (see Chapter 3). Such a comprehension highlights the multifaceted character of the region, and the Ruhr District now appears more diverse than the usual clichés of it. Therefore, its diversity is presented not as an entity of itself, but as a construction within which the individual elements stand in relationship to each other. Here, vitality opposes permanence, and the notion of a metropolis arises of something that is 'still in the making'.

However, the image of a 'dialogue of differences' poses a number of questions to which there are no clear answers: Who is conducting this dialogue? Why should it be assumed to be productive and not destructive? What conditions need to be in place for such a dialogue to be successful? What identity can emerge out of this 'dynamic structure'? Furthermore: does the underlying intention confirm existing stereotypes and clichés, whilst in the same breathe utilizing (high) cultural elements? To tackle these questions we have to go deeper into the region and reflect its postmodern experience.

Anyone visiting the new cultural tourism sites of the Ruhr District, such as for example the Zollverein Coal Mine, a World Heritage Site, or the Jahrhunderthalle in Bochum will inevitably also gain an impression of the surrounding areas. Both of these locations are located on sites that were directly tied to the old coal and steel industries, and have now been regenerated at great expense as highly prestigious cultural attractions. In terms of urban development, their surroundings certainly display deficits; moreover, they are both situated in neighbourhoods inhabited by a social class that is normally not envisaged as the target audience for cultural venues of this type. But if these cultural venues are to be successful over a longer term, it is essential, according to Klaus Kunzmann, that the areas surrounding the attractions are significantly enhanced in terms of urban quality (Kunzmann 2007: 7) and that their environment becomes an integral part of the site. Only then will the neighbourhoods be in a position to benefit economically and socially from the visitors that come into the area, and only then will an integrated social and cultural atmosphere arise.

The fact that it is simply unavoidable that visitors will permanently encounter the surrounding areas of the attractions may well explain why discrepancies cannot be denied. Such a circumstance goes in favour of the organizers' intentions to develop the *Year of the Capital of Culture* as something for everybody, and not only as an elitist concept of culture. But in order to build up a strong image which is capable of integrating a regional brand that surpasses social boundaries (as is the intention behind the label Ruhr Metropolis) it is vital to underline the region's overall metropolitan structure of difference, rather than focusing on inner-regional differences.

Such a construction of an image finds its parallels in the formation process of the inhabitants' personal identity. Here, it is also very significant that the regional identity is demarcated from the external world, thus keeping the past present in the social rooting of the people by memorizing the integration of the region in the process of German nation building in the late nineteenth century, through the industrialization which caused fundamental social changes in German society as a whole. The motto for Ruhr.2010 of 'Transformation through Culture – Culture through Transformation' describes this process of perceptive change from social modification to social identity building. This is precisely one of the points which should be highlighted in the image- and brand-creation activities of towns and regions: according to Ole B. Jensen, branding 'should take stock of prevailing local identities' (Jensen 2006: 87), but we insist that the actual identity building process is historically bound through experiences of the past, while it is now constructed by active social integration. In both cases, however, the process sets in motion what is locally or regionally specific, bringing together into one convincing narrative such diverse elements as myths, social history, monumentalizations, political ideologies, and feelings. What is meant by 'convincing' here is that it is comprehensible to and can be scrutinized by all of its recipients. As Jensen concludes: 'Seen in this light, urban branding is a sort of evocative story-telling aimed at "teaching" its recipients to see the city in a particular way' (ibid.: 86). In our opinion, this definition of urban branding describes precisely what those responsible for Ruhr.2010 have aspired to achieve: condensing the *Year of the Capital of Culture* into one story that combines most of the chapters of the region's history, putting its importance into the European context, and pointing to its post-industrial future.

Emotion Epitomized

The 'Capital of Culture' of the Ruhr District is mainly intended and justified as a political means to strengthen urban development, social policies and the economy. Ruhr.2010 is defended therefore on both planning and economic grounds. However, in the environment of postmodern virtual policies, the organizers are not relying solely on an appeal to the minds, but are also attempting to reach out to the hearts. This is, however, not a matter of addressing personal feelings, but of touching on atmospheric sensations that are charged with collective emotion: 'It was Albert Camus who stated that the origin of every piece of art lies in a simple and deep emotion, and it is this concept that characterizes the combined driving force behind the Capital of Culture. The Ruhr region is emotion epitomized. Emotion is the special ingredient that turns the story we are telling into something special' (Ruhr.2010 Gesellschaft 2009b: 5). Invoking existentialist Camus in such an emotional sense implies that the very interpretation of the organizers of the *Year of the European Capital of Culture* is to be seen as a creative process in itself – as art, and not simply as an umbrella for political action. What makes Ruhr.2010 unique, again according to the organizers, is the very special 'Ruhr feeling', a matter of collective experience that sustains the *ästhetische Arbeit* [aesthetic work] (Böhme). This emotional experience renders the culture-based transformation of the Ruhr District quite distinctive, constituting a special ingredient to Ruhr.2010.

Consequently, the three guiding themes of the 'Capital of Culture' appear as a fusion of narrative, image and social change, mingled in a general emotional regional atmosphere: 'The Ruhr mythology', 'Re-designing the metropolis' and 'Moving Europe' (Ruhr.2010 Gesellschaft 2009b: 5) Here, fine details and deeper understandings are relegated to institutionalized individual programme components *a posteriori*. Since emotions have been sparked by Ruhr.2010, the atmosphere is neither entirely clear nor only positive or negative (as shown above), and the actual events and their locations are participating in the process of emotional coding. Considered in these terms, Ruhr.2010 develops its own distinctive rhetoric through the tension between the organizers' ideas, the inhabitants' identification with the region and the tourists' images; for sure, this is not simply a process of random effects, but it is sustained by a diffuse combination of organized local attitudes, and tourism agents.

In January 2010, at the beginning of the *Year of the European Capital of Culture*, a total of 1000 people living in North Rhine-Westphalia were surveyed (Infratest dimap 2010) and asked whether they thought of the Ruhr District as a metropolis. 72 per cent responded positively, divided between 38 per cent saying 'very much so', and 34 per cent 'a little', thus demonstrating that their attitudes refer more to a gradual feeling, and less to a mental definition. It also becomes clear that there is a sense of place for the region as a metropolis, even if it is the notion of a 'still developing metropolis'. However, it should be emphasized that the survey revealed an important regional difference: while in the Ruhr District 81 per cent responded with 'yes', it was only 67 per cent in the rest of North-Rhine-Westphalia. This allows the conclusion that the perception of the Ruhr District as a metropolis is much more an insiders' view and a question of regional identity than an outsider experience. Consequently, the actual process of cultural imaging as a Capital of Culture passes through emotional components of intercultural encounters and must be interpreted as an attempt to overcome the deep-rooted sense of inferiority that arises from structural change accompanied by the remainder of the negative image of the region.

Conclusions

For the Ruhr Regional Association (RVR), as the principal shareholder of the Ruhr.2010 company, the *Year of the European Capital of Culture* is an important instrument for both marketing and regionalization. The region can now present itself as a united cultural entity, which in turn enhances the RVR's own legitimacy. Into this framework, the municipalities of the region position themselves through individual projects, which are subordinated to the general image by enhancing their own profiles vis-à-vis neighbouring towns and cities. Thus, Ruhr.2010 is perceived as a regional alliance with a common purpose, where various actors are united through their desire to improve the region's image and social situation.

In this context, culture has taken a decisive role as an instrument in forming the Ruhr District's identity, something which the region has never experienced before to this extent. However, in postmodern times it appears nearly as a 'panacea' (Hiß 2009: 10). With this policy, Ruhr.2010 has embarked on a programme of visual and atmospheric 'staging' aimed at giving the region a 'more metropolitan' identity to create attractiveness. An essential component for this identity policy is memory. However, few elements of a critical perception of the region's history and memory are included (ibid.: 12); it is much more the positive 'staging' of touristic attractiveness of the region, and less that of social attractiveness through identity building, that highlights the experience of people within the region. But, as the experience of this urban space is tinged by deficits in terms of both structure and built development, the 'real' situation is in sharp contrast to showcasing the region's industrial heritage for tourists. And it is exactly this experience of social discrepancy that has always permeated local aesthetics. Thus, the resulting emotional tension and ambiguity even today still dramatizes the region's social history and geography through tensions, and defines its attraction in contrast to mostly 'clean' consumers' atmospheres. The aesthetic experience influences not only visitors' attitudes to the region, but also the strategy adopted by the organizers, who have tried nonetheless to exploit this discrepancy positively as a contrast.

From this perspective, it is more than logical that the merely infrastructural thought system of creating visitors' hubs to assist the tourists' sense of orientation is not at all consistent with the way in which local residents perceived their own territory. This strategy of the planners divided the experiences of both groups. However, when the emotional rooting through vision was implemented through panoramic views, this allowed a common ground both for local residents and visitors. From here, they could share the experience of the region, positioning themselves 'on top' of the region and creating a single entity from above. Thus, both groups have been fully accepted as perceptive actors. It is this construing the region through a tourist's perspective, through a tourist gaze, that now also permeates the local view.

When the question was transferred to the envisioned goal to present the region as a metropolis, however, the visual approach failed. A set of associations commonly evoked by a typical metropolis did not coincide with the images of the Ruhr metropolis. Then, only construing a distinct atmosphere of metropolis allowed a going beyond typical images, and again a common (and very innovative) ground has now been found both for visitors and inhabitants, in an atmospheric sense. Such a deliberately produced atmosphere of a 'metropolis' is now open for an emotional binding of community, between visitors (and outsiders in general) and local residents, and allows both groups to experience what makes the region distinctive and unique. Here, the image of a 'Capital of Culture' is founded on the thesis that the region is emotion epitomized and that the identification with the region can first be felt by those who are living there, but can also be transmitted and made tangible for those who come from outside. This strategy of 'staging' a corresponding atmosphere between both groups represents a mechanism where the social integration via images and aesthetical work is more than an adornment in the planning process. In postmodern atmospheres of a consumer society, the creation of attraction is not only a proliferation of postmodern capitalist attitudes, but also a political means to forge new processes of social integration. The attempt to dispel the impression of 'provinciality' (and thus submission to hegemonic powers) is only possible via self-assumed identity, expressed in images and atmospheres. Therefore, the Ruhr.2010 project is an instructive example of how attraction and fascination can become (critical) social forces that forge a postmodern society, even if they are embedded into the capitalist tourism complex. This makes the formerly depreciated Ruhr District a prototype for cultural restructuring.

References

Betz, G. 2008. 'Von der Idee zum Titelträger. Regionale Kooperationsprozesse des Ruhrgebiets bei der Bewerbung zur Kulturhauptstadt Europas 2010', in *Die Idee der Kulturhauptstadt Europas. Anfänge, Ausgestaltung und Auswirkungen europäischer Kulturpolitik*, edited by J. Mittag. Essen: Klartext, 191–213.
Böhme, G. 1995. *Atmosphäre. Essays zur neuen Ästhetik*. Frankfurt am Main: Suhrkamp.
———— 1998. *Anmutungen. Über das Atmosphärische*. Ostfildern: Ed. Tertium.

de Michelis, M. 2006. 'The Little City', in *M City. European Cityscapes. Exhibition catalogue, Kunsthaus Graz*, edited by M. de Michelis and P. Pakesch. Cologne: Walther König, 10–27.

European Commission 2010. *The Present, Future and Past Capitals*. [Online]. Available at: http://ec.europa.eu/culture/our-programmes-and-actions/doc481_ en.htm [accessed: 7 February 2010].

Ferguson, F. 2006. 'Where Density Ends. Conversation with Susanne Hauser', in *Talking Cities. The Micropolitics of Urban Space*, edited by F. Ferguson. Basel: Birkhäuser, 42–45.

Groys, B. 1997. 'Die Stadt auf Durchreise', in *Logik der Sammlung. Am Ende des musealen Zeitalters*. München: Carl Hanser, 92–108.

Hauser, S. 1990. *Der Blick auf die Stadt. Semiotische Untersuchungen zur literarischen Wahrnehmung bis 1910*. Historische Anthropologie, vol. 12. Berlin: Reimer.

Hiß, G. 2009. '"Kultur als Klammer". Zur Inszenierung einer Metropole', in *Schauplatz Ruhr. Jahrbuch zum Theater im Ruhrgebiet. RUHR.2010 – Inszenierung einer Metropole*. Berlin: Theater der Zeit, 8–12.

Infratest dimap 2010. *Kulturhauptstadt RUHR.2010: Bekanntheit in Deutschland und NRW*. [Online]. Available at: www.infratest-dimap.de/umfragen-analysen/ bundesweit/umfragen/aktuell/kulturhauptstadt-ruhr2010-bekanntheit-in-deutschland-und-nrw/ [accessed: 07 February 2010].

Jensen, O.B. 2006. 'Urban Branding, Image Formation and Regional Growth', in *The other Cities. Die anderen Städte: IBA Stadtumbau 2010. Urban distinctiveness*, vol. 4. Edition Bauhaus, vol. 22, edited by R. Sonnabend and R. Stein. Berlin: Jovis, 84–89.

Koolhaas, R. 2006. 'Nach dem Iconic Turn. Strategien zur Vermeidung architektonischer Ikonen', in *Iconic Worlds. Neue Bilderwelten und Wissensräume*, edited by C. Maar and H. Burda. Cologne: Dumont, 107–29.

Koschorke, A. 1990. *Die Geschichte des Horizonts. Grenze und Grenzüberschreitung in literarischen Landschaftsbildern*. Frankfurt am Main: Suhrkamp.

Kunzmann, K.R. 2007. 'Kulturhauptstadt Essen: eine Chance!'. *RaumPlanung*, 130(1), 5–10.

Lootsma, B. 2006. 'Ruhr City – A City that Is, Will Be or Has Been', in *M City. European Cityscapes. Exhibition catalogue, Kunsthaus Graz*, edited by M. de Michelis and P. Pakesch. Cologne: Walther König, 194–202.

Pagenstecher, C. 2003. *Der bundesdeutsche Tourismus. Ansätze zu einer Visual History: Urlaubsprospekte, Reiseführer, Fotoalben 1950–1990*. Studien zur Zeitgeschichte, vol. 34. Hamburg: Dr. Kovac.

Projekt Ruhr Gesellschaft. 2005. *Masterplan Emscher Landschaftspark 2010*. Essen: Klartext.

Prossek, A. 2005. 'Zwischen Kitsch und Kathedralen: die Ruhrtriennale und das Ruhrgebiet', in *Themenorte*. Geographie, vol. 17, edited by M. Flitner and J. Lossau. Münster: LIT, 45–58.

———— 2008. 'Sie nennen es Metropole. Die Kulturregion Ruhrgebiet zwischen Anspruch und Wirklichkeit', in *Echt! Pop-Protokolle aus dem Ruhrgebiet*, edited by J. Springer, C. Steinbrink and C. Werthschulte. Duisburg: Salon Alter Hammer, 92–112.

Rehfeld, D. 2006. 'Profiling and Specialization of Old Industrial Cities – Trends and Concepts Based on the Example of the Ruhr Area', in *The Other Cities. Die anderen Städte. IBA Stadtumbau 2010*. Urban distinctiveness, vol. 4. Edition Bauhaus, vol. 22, edited by R. Sonnabend and R. Stein. Berlin: Jovis, 112–119.

Ruhr.2010 Gesellschaft 2008. *European Capital of Culture 2010 'Essen for the Ruhr'. Book one.* Essen: self-published.

———— 2009a. *Mission Statement*. [Online]. Available at: http://www.essen-fuer-das-ruhrgebiet.ruhr2010.de/en/organisation/unternehmenskultur/mission-statement.html" [accessed: 23 April 2009].

———— 2009b. *European Capital of Culture 2010 'Essen for the Ruhr'. Book two.* Essen: self-published.

Ruhrgebiet Tourismus Gesellschaft (ed.) 2007. *Kulturtouristisches Begleitkonzept Bewerbung 'Essen für das Ruhrgebiet – Kulturhauptstadt Europas 2010'.* Essen: self-published.

Salij, T.H. 2002. 'Observations', in *RheinRuhrCity – The Hidden Metropolis*, edited by NRW-Forum Kultur und Wissenschaft. Ostfildern-Ruit: Hatje-Cantz, 244–259.

Schmitt, P. 2007. *Raumpolitische Diskurse um Metropolregionen. Eine Spurensuche im Verdichtungsraum Rhein-Ruhr.* Metropolis und Region, vol. 1. Dortmund: Rohn.

Selle, K. 2005. *Planen. Steuern. Entwickeln. Über den Beitrag öffentlicher Akteure zur Entwicklung von Stadt und Land.* Dortmund: Dortmunder Vertrieb für Bau- und Planungsliteratur.

Urry, J. 1992. 'The Tourist Gaze and the 'Environment'. *Theory, Culture and Society*, 9(3), 1–26.

———— 2002. *The Tourist Gaze.* London: Sage Publications.

Wachten, K. (ed.) 1996. *Change without Growth? Wandel ohne Wachstum? Stadt-Bau-Kultur im 21. Jahrhundert.* Braunschweig, Wiesbaden: Vieweg and Teubner.

Wagner, B. 1988. 'Vom Aschenputtel zum Hätschelkind? Tendenzen kommunaler Kulturpolitik', in *Kultur macht Politik. Wie mit Kultur Stadt/Staat zu machen ist*, edited by A.R.T. Cologne: Kölner Volksblatt-Verlag, 68–94.

Zukin, S. 1995. *The Cultures of Cities.* Cambridge: Blackwell Publishers.

Chapter 10

Strategic Staging of Urbanity: Urban Images in Films and Film Images in Hamburg's City Marketing

Sybille Bauriedl, Anke Strüver

Geographical Imagination of Urbanity – Representation of Urban Landscapes

More than a decade ago, David Harvey observed that the contemporary restructuring of urban spaces has been accompanied by changing economic and cultural practices in cities (Harvey 1990). In this context, Sharon Zukin (1995) noticed a symbolic economy of cities shaped by those who want to achieve financial benefit or advantages in the competition of cities. In a similar vein, Lash and Urry (1994: 64) have stated: 'Economic and symbolic processes are more than ever interlaced and interarticulated ... The economy is increasingly culturally inflected and ... culture is more and more economically inflected' (see also du Gay and Pryke 2002 for the culturalization of the economic sphere and the economization of the cultural). And Mark Gottdiener declared that the production of signs is elementary for entrepreneurial success, because symbolic value seems more important for economic success than commodity value (Gottdiener 2001). As a result of these accounts urban scholars should concentrate on the development of empirical instruments which make it possible to analyze symbolic processes as part of the urban condition. In this sense – and in order to understand the economic and symbolic, as well as discursive processes – we have to look at both the sphere of production and the sphere of consumption of themed urban landscapes, since both the production and consumption of urban places and spaces are actively constructed by economic and political, as well as social and cultural, practices.

By discussing the interrelation of these spheres, this contribution focuses on the production and consumption of signs – of distinctive iconographic signifiers – in films and city marketing. Visual images shape our understanding of, and reaction to, the world we live in – and this world is a cinematic world. 'The contemporary society knows itself unreflexively, only through the reflections that flow from the camera's eye' (Baudrillard in AlSayyad 2006: 2). Film images are so powerful for geographical imaginations that they are increasingly produced and used in city marketing strategies. To represent and interpret urbanity (and urban landscapes of postmodernity), urban politics are inspired by, and orientated towards, film images of urbanity. This means that the production of films with locations in distinctive cities produces urban places and spaces at the same time.

In order to approach urban images in films and film images in urban development, we accept that '[f]ilm geography is not simply a disassociated "reading" of entertaining "texts", but rather, are inquiries of cultural documents that reveal hegemonic tensions within meaning creation, appropriation and contestation' (Lukinbeal and Zimmermann 2006: 318). However, these hegemonic tensions include, on the one hand, a production of meaning(s) beyond the simple act of signification, i.e. of ascribing meanings. They refer to the *production of fascination* in an aesthetical and emotional sense as well as in a politically contested way. On the other hand, political contestation refers to a dimension of the production of meaning and fascination that touches both the level of urban government and the micro-scale of everyday life and urban governance.

Against this background, this chapter focuses on two perspectives within film geography: the first, cultural studies (and its post-structuralist background), positions film as an arena in which socio-spatial meanings are defined, perceived and experienced, but also contested and negotiated. The second, a political economy perspective, discusses the implementation of urban images created in films for city marketing. Consequently, we will discuss visual representations of urbanity (film images of urban space, resulting in urban imaginations) and the 'strategic staging of urbanity' (urban imaginations drawn from films in urban development policies and place-promotion strategies).

'The new city is characterized by "entrepreneurial governance" and by vigorous programmes of place promotion, and is also visibly more spectacular, with a revitalized city centre featuring gleaming high-rise office blocks, iconic "flagship" public buildings, waterfront developments, heritage centres and urban villages. Well-known examples include London's Canary Wharf, Barcelona's Olympic Marina, Paris' La Defense, Vancouver's Pacific Plaza, New York's Battery Park, Cardiff Bay and Pyrmont-Ultimo on Sydney waterfront' (Jayne 2006: 39) – in addition to Hamburg's HafenCity, which we will employ as an example below.

First, we will introduce our methodological background and argue that distinctive urban spaces are used as sceneries for cinematic staging and at the same time for the strategic staging of urban politics. Then we will move on to the ascription of meanings by city marketing and the film industry and analyze how they enforce each other and organize a 're-cycling' of urban images. Both parts, however, rely on the ideas that (1) the production of fascination is not reduced to the aesthetical and/or emotional sphere, but is inherently political and that (2) this form of fascination is also dominant in people's everyday life.

Theoretical Setting: The Interaction of Urban Images in Films and Film Images in Urban Development Strategies

In order to investigate the mutual relations of urban images in films and film images in city marketing strategies we adopt a methodology that takes the changing functions of urban places and spaces into account and focuses on the influence of film representations on urban imaginations. It is thus less concerned with the city itself, but rather with the construction of urban realities through visual media such as film – particularly with the 're-cycling' of film images and related imaginations in city-marketing strategies.

Generally speaking, visual media and representations such as film images offer particular visions of the world. Yet they do not simply depict, but actually make and shape the world in visual and narrative terms. It is well-known that those representations are not neutral or objective but rather forms of discursive practices. Nevertheless, it is often assumed that what is *media-ted* as 'real' *is* factual. Media representations in general are thus not mere carriers of information, rather they play a crucial role in producing interpretations of representations: They are constitutive of meanings and they are 'a *means* for the geographer to intervene in the production of the "real"' (Craine 2007: 148, original emphasis, see also, for example, Rose 2001, Hall 1997, Duncan and Ley 1993).

In the present context, 'representations' encompass two aspects, 'presentations-and-their-interpretations' such as film narratives and images and the meanings they create, on the one hand, and imaginations as imagined realities on the other. Consequently this is a methodology that is based on Roland Barthes' claim of 'imagination as signification' (1982), which accepts the impact of images on imagined realities, but also on strategic urban(-development) image-campaigns.

Against this background and because of our concentration on the discursive production of urban spaces and images – including the re-cycling of these images in marketing campaigns – the chapter relies on a post-structuralist approach that understands meaning as being produced through language, i.e. in all sorts of 'texts', including visual images. Although there is no agreement upon a single methodological approach within poststructuralism, it is possible to distinguish between two broad methodological streams, namely the semiotic and the discursive: The semiotic approach refers to semiotics as the study of signs and their general role as *carrier* of meanings, focusing on *how* language works and on *how* meanings are derived from narratives and images (Culler 1976, de Saussure 1960). The discursive approach, on the other hand, refers to discourses as ways of constructing knowledge about a particular topic. It is about the formation of ideas, images and practices, which provide ways of talking, forms of knowledge and are conducted within particular social activities and institutions. The discursive approach is more concerned about the *effects and consequences* of representations and can be conceptualized as *complementing the semiotic approach* by making it possible to analyze the effects which these meanings have (see also Waitt 2005, Gibson-Graham 2000, Foucault 1973). In dealing with the production and perception of urban images based on film elements and concentrating on their meanings and effects, we argue in the following that a combination of the semiotic and the discursive approaches is useful in order to cover both the process through which meanings are assigned to urban places as well as the effects of such imaginations (imaginations understood as outlined above: as signification and 'imagined realities').

Semiotics – the study of how meanings are transmitted and understood – represents a methodology for the analysis of 'texts', i.e. any form of message/ communication, including visual ones. Semiotic approaches seek to reveal different levels of meaning, including hidden meanings and motivations in, for example, the visual media and advertising. What is more, they even argue that the only way to understand the world is through 'texts', such as visual and narrative representations. A semiotic analysis is thus one way to approach the question of *how and which meanings – such as the idea of something forming a 'city' – are derived from film narratives and images.*

Roland Barthes (1967, 1973, 1977) has provided a very useful exposition of semiotics in relation to popular culture, which can be applied to film images. Semiotics uses different analytical tools: Traditionally, semiotics has always paid attention to the representation itself. More recently, however, the field of 'social semiotics' has pointed out that to understand a representation it is also necessary to take its audience and thus its reception into account. Stuart Hall was among the early critics of empiricist and behaviorist explanations of representations: Rather than understanding the process of media-communication as a direct process extending from sender to receiver (e.g. movies and their audiences), he underscored the differences that often exist in the producers' and the audiences' interpretation of a representation – between its production (encoding, e.g. the production of a movie and its 'intended message(s)') and its reception (decoding, e.g. the meanings given to it by people who consume it) (Hall 1980, 1997). Audiences therefore need to be understood as 'co-authors' of a representation, i.e. not as passive recipients, but as producers of alternative interpretations of images and alternative *imaginations*. A critical analysis of visual material thus approaches viewers as *active users* of this material. Michel de Certeau, for example, has stated that discourses 'are marked by *uses*; they offer to analysis the imprints of acts or of processes of enunciation ..., they thus indicate a social historicity in which systems of representations or processes of fabrication no longer appear only as normative frameworks but also as tools manipulated by users' (de Certeau 1984: 21, emphasis added). He found out that the circulation of a representation does not reveal much about its perception, its 'use' by readers. Therefore he suggests that the 'difference or similarity between the production of the image and the secondary production hidden in the process of its utilization' be investigated (ibid.: xiii, for (geographical) applications, see Strüver 2005). A user's making is another production, but one that manifests itself through the use of products. This is why de Certeau advises to analyze what users make of the representations they encounter. And this is also true for 'audiences' and 'users' in the sense of city-marketing-organizations and -strategies: a perspective we will draw on below.

It is important not only for this new perspective on 'audiences' but in general to acknowledge that although a representation's meanings are complex and multiple (polysemic), they are not free-floating, but should be contextualized. This contextualization in our sense goes beyond the ('embedded') meaning in its semiotic sense, and involves its discursive function, which is based on the attempt to create emotional links. Moreover, representations and their meanings do not simply exist in people's minds. They have real effects through their function of setting values and norms and they are mobilized through ways of 'seeing' and 'reading'. According to Barthes (1977: 19), an image 'is not only perceived, received, it is read, connected more or less consciously by the public that consumes it' to an already existing stock of signs. In broadening the latter by focusing on the *use* of images and their meanings in city-marketing-strategies and in addition to the mere *production* of meanings it is necessary to take the effects of these meanings – and their strategic application – into account. Here, it is possible to draw on Foucault's discourse theory, which is concerned about the ways in which meanings get legitimized, normalized and finally accepted as reality. Discourse theory concentrates on the *effects and consequences* of the meanings produced – and fascination is one of these effects.

'Discourse', in Foucault's framework, is about (various) practices that produce, maintain and sometimes transform meanings – all of them embedded in power relations and knowledge systems, which can also be contested, on both the author's and the reader's side (Foucault 1980). Representation for Foucault (1973) is a discursive practice, presupposing a 'regime of truth' but in fact, there are no *neutral* or 'original' representations.

An analysis of discourses thus explores the outcomes of discourses on actions and perceptions. It aims at identifying the regulatory frameworks in which discourses are produced, circulated and communicated as well as uncovering the internal mechanisms that maintain 'common-sense' (Waitt 2005). Against this background, it seems appropriate with this approach (similar to semiotics) to study what effects geographical imaginations have on urban development strategies. The question at hand then is which film images result in a 're-cycling' – or 'use' – of these images (and their perceived meanings) for marketing campaigns in the context of place promotion? And if urban images are a tool to create fascination – and if fascination functions here as a link between the semiotic and the discursive approach – the question is also how these urban development strategies produce emotional effects.

According to Ford (1994), cities are perceived via the role they 'inhabit' in films. Their role has thus changed from being a background or setting for the narrative to being part of the narrative. And since place and space are not passive backgrounds – neither for any kind of social encounter, nor for visual representations such as film images and narratives – their investigation rests on the complex interactions between economic, social, political and cultural practices.

As noted above, the emerging field of 'film geography' thus goes beyond the analyses of films as entertaining visual texts; thus (1) it is also concerned about films as cultural documents, which actively produce (contested) meanings and (2) asks whether the entertaining effects attempt to introduce meanings subversively, which later (re-)configure the images of the spectators as guidelines in their urban everyday life. In general, there are three dominant perspectives within film geography, namely the geography *in* film, the geography *of* film and the geography *from* film (see Bollhöfer 2007, Bollhöfer and Strüver 2005, Hopkins 1994) – all of them relying on and supporting the idea that films deal – and play – with place and space, but also construct place and space and their meanings while dealing with them. Aitken and Dixon (2006: 327f.), on the other hand, refer to and broaden these three perspectives in a way that focuses on how meanings are ascribed to people and place via their appearances on screen, how these meanings intersect with dominant meanings produced by other popular media and on the interplay of film-technologies and (embodied) perceptions. At the same time, they point out that 'geographic concern was also lacking a critical perspective, focusing primarily on articulating how certain films portrayed a quirky geographic realism rather than more pithy issues of how they produced meanings' (Aitken and Dixon 2006: 328). Accepting this critique and the distinctive perspectives within film geography, we analyze the production, circulation and adoption of these images, not to say *the production of the effect of fascination*, for city marketing strategies.

Urban Images in Film – Film Images of Urbanity

The city is not only what appears on the screen, but it is also the image of urbanity made by the medium of cinema (AlSayyad 2006: 2). Visualizing cities and urban images has been one of the essentials of filmmaking since 1890. The starting point of cinematic image production was in the period of urbanization in the context of an expansion of industrial capitalism in Europe and the USA. The cinema especially represented a modern lifestyle and the urban industrial condition of this time. The Modern City is characterized by movements and the feature and documentary films record these movements. The past hundred years of urban image production in films create the feeling of walking within a film scene in many European and North-American cities (see Clarke 1997). The mutual relations between cinema, modernism and urban development are well discussed by cultural scientists (e.g. Donald 1999, Clarke 1997, Aitken and Zonn 1994) and philosophers (e.g. Baudrillard 2004, Benjamin 1980) with references to a set of films such as *Berlin: Symphony of a Big City* (1927), *Metropolis* (1927) or *Modern Times* (1936).

Movies usually typecast events in a binary fashion, such as dualisms of nature-culture, environment-economics, science-politics etc. This logic is one way films naturalize cultural politics. But these binary categories which characterize the urban condition in cinema are already socially constructed and produced by power relations. Therefore, the cinematic city is constructed by urban images based on dualisms. From the 1920s until the 1970s the dualism between working class urban life worlds and post industrial modern ways of life was the usual plot of urban narratives. For example, the films of the French film maker Jacques Tati *Playtime* (1967) and *Mon Oncle* [*My Uncle*] (1958) represent this dualism in an explicit way. Here, he presents architecture, street life and industrial and post industrial work places as symbols of a post-Fordist transition. His films thus exemplify the image of a city as a labyrinth of a dirty nineteenth century urban landscape and the coexistent image of a city as synthetic modernist urban place. *Mon Oncle* tells the story of the clash of these two worlds while the leading actor wanders between a traditional city block and a modern cubist villa to critique the marriage between urban renewal as a policy and modernism as an architectural ideology. Both films 'focus directly on disillusion with the physical outcome of modern architecture and urbanism' (AlSayyad 2006: 9).

The transition of urbanity in the context of postmodernity is recognized as an essential element of films, which represents forms of simulation and hyperreality as urban attributes. Until the mid 1970s urban postmodernity was accompanied by new types of economic arrangements with changing patterns of labour, production and consumption and was represented in films as a crisis of urban life. Big cities were depicted as places of social and emotional degeneration (as in *Blade Runner* from 1982 or *Manhattan* from 1979), turned in a dystopic hell (like in *Falling Down* from 1993) or as the epicentre of the man made ecological crisis (as in *The Day After Tomorrow* 2004). 'The postmodern condition then only appears dystopic when contrasted with the idealized modern experience' (AlSayyad 2006: 123). Films like *Truman Show* (1998) represent the utopian alternative of the postmodern urban condition of fragmentation, decentralization and vulnerability. The dualistic representation of metropolis as dystopic hell, and small cities as idyllic heaven was an image invoked in many Hollywood productions. In *Falling Down* we see the urban landscape of 'Los Angeles constructed as fragmented, hostile, violent, unreadable – and therefore out of control' (AlSayyad 2006: 11). In contrast the movie *The Truman Show* portrays a city with happy residents in a commercialized, manipulated and controlled utopia; nevertheless in the end this cinematic city is exposed as a claustrophobic fake town.

According to AlSayyad (2006), the present postmodernist fragmentation in films cannot be understood without the embedded assumption of modernist desire. The film images of cinematic urbanity today represent a fusion of modern and postmodern elements of urban condition and landscape. The common practices of producing cinematic urbanity use creative narratives and fragmented realities to unsettle stable beliefs of modernity and postmodernity and break down the boundaries between the real and the reel city. The best known cinematic city of the postmodern condition is Los Angeles. We all know the economic and spatial manifestations of fragmentations by class and ethnic distinctions and we know the vulnerability of the city to ecological disasters such as earthquakes and fires in Los Angeles, even though we have never been to nor have ever seen the 'real' city.

Because of the discursive links between the cinematic city and the debate on urbanity, we think it is useful to raise film to its proper status as an analytical tool for analysing urban discourses and conditions: What images and imaginations are produced and reproduced in order to 'theme' urban landscapes in film and in the debate on urbanity? What are the geographical imaginations of the cinematic city? The cinematic representations of modern and postmodern urban conditions in the Hollywood films mentioned above are part of the repertoire of signs and knowledge systems based on cultural practices in Western societies. They are the references for creating urbanity both in recent films and in city marketing strategies.

We will apply this discursive relation to the example of Hamburg as cinematic city. We argue that the postindustrial metropolis creates its fascination out of this intersection of reel images and the ideals of post-Fordist urbanity. Cities are more and more identified by urban images which are intelligible beyond the local context. Even if the material condition is the origin of urban images, reel images are an important asset for creating fascination and for persisting in the international competition of economies of fascinating cities.

Objectives of City Marketing and Film Industry in Hamburg

Hamburg is a good example to illustrate our argument, since the city is an important location for international and national film production. Its city council acts within the ideological frame of an entrepreneurial city and intensely promotes city marketing strategies. In this respect, Hamburg represents a specific cinematic city and its city marketing strategies partly rely on film images that create visual icons for competition strategies with other cities. In the following section, we discuss city marketing in the sense of locational management which understands the city as a result of image production and pushes the objectives of enhancing the attractiveness of the city as business location, place of consumption and destination for tourists.

Hamburg has a long tradition as a commercial city and port, but since the 1970s its economy is increasingly shaped by harbour related services such as insurance, banking and other services. Many companies in these branches, as well as media companies, have their headquarters in the city centre. The city's urban policy is organized along the regime of economic and demographic growth, with the objective to increase the economic power of the city and the quality of life for its inhabitants. The waterfront of the inner city is the major urban space in this context for attracting new residents, service businesses and tourists.

Numerous national and regional television stations – also producing films (*ARD, NDR, Studio Hamburg*) – and more than fifty private film companies are located in Hamburg. A federal film promotion agency was founded in 1979, which has the objective to promote the appearance of selected city locations in feature films and TV series. It has an annual budget of seven million Euros a year. The film locations most often chosen in Hamburg are the inner city waterfront, the red

Figure 10.1   Film still of *Die Entscheidung* [*The Decision*], Germany 2005
Source: Studio Hamburg Produktion GmbH

light district 'Reeperbahn' in St. Pauli, the various squares and bridges, and Lake Alster in the city centre. These are film locations which can be easily recognized by national and international audiences.

Film locations at the waterfront are increasingly being shot in feature films and TV series. There are actually two police series, one emergency medical service series and two lawyer series weekly on TV with waterfront locations in each episode. The most important film locations are sites at the former harbour on the northern bank of the River Elbe, new office buildings at the restructured waterfront and the recently built HafenCity in the former docklands. The two following figures are examples of film locations sponsored by the film promotion agency Hamburg/Schleswig-Holstein in recent films shot exclusively in Hamburg.

The TV feature film *Die Entscheidung* [*The Decision*] narrates a crime story in which the main character has emotional troubles. The main film location is one of the most recently completed office buildings on the restructured waterfront called 'dockland'. Through the window of this office one can see the container terminals of the harbour on the opposite bank of the River Elbe in the background. The blue collar working space in the background is used as a backdrop for white collar workers and creates an iconographic symbol for the new condition of post-Fordist urban lifestyle. This film image represents the spatial link between new economy and old economy. With the distance across the river the harbour becomes an aesthetic scene and turns into a setting for new post-Fordist businesses. The

Figure 10.2   Set photo of *Paulas Geheimnis* [*Paula's Secret*], Germany 2006
Source: Element E Filmproduktion

marketing of these offices for service sector companies refers to the emotional attitude of being a successful 'captain' of the business. Especially at the waterfront, real estate and marketing agents use emotional reference to maritime attributes to attract potential costumers. They evoke positive associations and romanticized ideas of seafaring and a traditional harbour economy to commercialize office locations on the waterfront.

The movie *Paulas Geheimnis* (a story of two children hunting a criminal in the city of Hamburg) is also set in the HafenCity. Because of its location in the new urban quarter of Hamburg, the film refers to a postmodern image of urban landscapes and harbour relics on the waterfront.

Encoding and Decoding Locations at the Waterfront

'In the construction of meaning attached to urban and natural settings, the primary purpose of the [film] image is *not* that of place promotion. Nor is the main purpose to construct an accurate image of urban or natural settings. The primary purpose is to sell a story and to entertain. The images used in film are created and selected based on their aesthetic qualities, entertainment value and ability to strengthen the story' (Kennedy and Lukinbeal 1997: 42, emphasis added).

Taking this quote as a 'point of departure', we will now put more emphasis on the 're-cycling' of film images and related imaginations in city-marketing/ development strategies. We do not concentrate here only on how popular films are perceived and act as place promotion (see Ward 1998, Gold and Ward 1994), but also look at the works on film tourism (e.g. Bollhöfer and Strüver 2005, Zimmermann 2003, Katz 1999), and at how these places (and their developers) adopt films and film images *strategically for* their place promotion.

Gillian Rose (2001) applies visual methodologies to analyze the production of meanings investigating (1) the production of a representation, (2) the representation itself and (3) its audience(s). Such an approach can be combined with Hopkins (1994: 50, 61), who depicts a geography *in* film (depiction of spaces in films), a geography *of* film (imaginations of spaces constructed by film images) and a geography *from* film (reception of spatial imaginations).

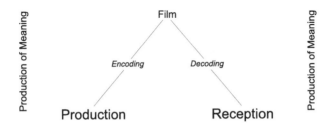

Figure 10.3   The production of meaning (1): interplay of film images and imaginations (revised adoption from Hall 1980)

We add a fourth element, namely an analysis of '*strategic* geographies *from* film', by concentrating on the strategic staging of urbanity by investigating urban images in films and how they are adopted for urban development strategies:

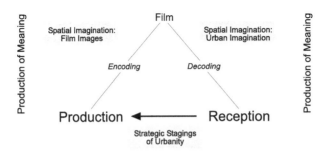

Figure 10.4   The production of meaning (2): interplay of film images and 'strategic re-cycling' of film images/urban imaginations

In this context, the new element for analysis ('strategic stagings of urbanity') concentrates on recycled film images and related imaginations in place-promotion-strategies, referring to an urban 'economy of signs' and an analysis of symbolic processes as part of the present urban condition. In doing so, we follow the general idea that films are not visual reflections of 'reality' (see above), but devices that produce cultural meanings and that transfer and circulate meanings.

An 'economic geography of film' thus focuses not only on locational aspects (see, e.g. Lütke 2006 for Cologne and Krätke 2002 for Potsdam-Babelsberg) or the profit of film industries and local economies' shares of these profits, but also on putting urban film images on the market in order to advertise a distinctive urban place.

In particular, we follow the analytical distinction between 'landscape as place' and 'landscape as space' (see Lukinbeal 2004, 2005). The first refers to a location's 'sense of place' that serves to *ground* the narrative of a film: It refers to a location where the narrative takes place (supposed realism), using iconographic signifiers to determine a filmic territory to a particular place, e.g. Hamburg. 'Landscape as space', on the other hand, refers to a 'placeless space' that serves just as background and is subordinated to the film narrative. Transferred to our empirical example from Hamburg, the metropolitan area of Hamburg functions as a distinctive *place* with unique features, whereas the HafenCity acts merely as a placeless space (e.g. office blocks along the waterfront). The question related to the 'urban economies of fascination' then, is whether here

- 'place is turned into space': i.e. whether the images and imagination of Hamburg are complementary and overlap with those of the HafenCity and inner city waterfront, in a way that Hamburg tends to 'vanish' into HafenCity or vice versa
- 'space is turned into place': i.e. whether the images and imagination of the HafenCity and waterfront are so dominant that they serve as a *local* sign and advertisement for Hamburg, but in a *global* architecture and structure of waterfront development

It seems that fascination becomes involved in an ambiguous relationship between place and space where the clearly differentiated separation between both of them tends to be dissolved.

Reception of Film Images: Examples from Hamburg for the Strategic Staging of Film Images

Meaning is naturalized through the infinite and obsessive repetition of images and narratives: 'it must be "true" if we see it everywhere all the time' (Lukinbeal and Zimmermann 2006: 315). The discursive practice of performance here is used strategically and in order to make specific arguments of urban policy relevant. Hence

**Figure 10.5   Film images in city marketing publication (detail), Hamburg
HafenCity development project**

Source: Hamburg Marketing GmbH

the meaning of an image must be obvious – and understandable – for everyone to
create a themed place which carries the desired argument or activity (Flitner and
Lossau 2005). In this sense urban landscapes are always discursive landscapes.
They have to be analyzed within the discursive context of the knowledge of, and
practices in, urban spaces and places of a specific city.

In this sense, Hamburg's city marketing employs the imaginations of urban
life and urban landscape in films, invoking the images of the city for marketing
strategies. Publications of Hamburg's city marketing agency thus present
illustrations which refer to popular film images of urban areas and specific places.
One example is a recent publication of the HafenCity, where film images from the
above mentioned production *Die Entscheidung* are used for marketing.

The HafenCity as the most important urban development project in Hamburg
is presented here as a new urban district on the inner city waterfront. The buildings
and the public space at the waterfront are generic, and yet distinctive parts are
used for the locational strategy with the intention to produce a city of fascination.
Thus, the competitive urban policy aims to advertise Hamburg as an attractive
area for urban living and office locations, similar to Sydney, Boston, Barcelona
and Toronto. The HafenCity development project is accompanied by a debate on
new urbanity and the revitalization of European cities. In this context, Hamburg's

Figure 10.6   Film images in a city marketing publication (detail), 'Dockland'
office building
Source: Hamburg Marketing GmbH

city marketing presents the same images we know already from films that exposed Hamburg with a mixture of harbour relics and traditional water-front buildings on the one hand and postmodern architecture and attractive squares for various leisure and consumption activities on the other, thus transporting ideas of place and space simultaneously.

Another urban development project, the 'Elbe String of Pearls', has the objective to restructure the waterfront on the northern bank of the River Elbe for luxurious office and apartment buildings and fashionable restaurants. This objective is accompanied by an architectural re-design of the waterfront. The waterfront, which was notorious as being noisy and dirty until the 1980s, is presented now as an attractive living and working location. The most important attraction of the north bank waterfront, however, is the view of the active harbour area on the south bank, creating a real world film scenery, which makes a space into a place.

Both examples demonstrate that a film image of the waterfront is relevant as strategic scenery for producing fascination in different categories of urban landscapes: (1) the urban landscape of a traditional harbour city is used with relics of harbour industries in public space, (2) the urban landscape of a European city is created by density, functional mixture, open spaces and (3) the urban landscape of postmodernity is construed through urban spaces of a post-Fordist society with squares and boulevards for commercialized recreation and tourism, embedded into a postmodern architecture. Here, specific forms of urbanity have become important criteria in competition between international cities. Signs of urbanity, like a creative urban lifestyle with a 24/7 time regime, clean and safe open and green spaces, as well as a tolerant atmosphere for people with various cultural backgrounds and sexual orientations, become more and more important as locational factors. These are signs of a symbolic economy of urban development. Within this strategy of an entrepreneurial city we can recognize a harmonization of urban identity and city marketing through the use of a globalized sign system (Mattissek 2008). These practices of urban politics are characterized by a market economy ideology that concentrates on a corporate identity for global commercialization, and here the limits between space and place are eroded.

The increasing economy of fascination in urban politics is increasingly influencing the process of realizing megaprojects in Hamburg, like the Hamburg HafenCity or the 'Elbe string of pearls'. Thus, the debate on local urbanity and the staging of urban images is an instrument for the international representation of a city. Within this symbolic economy of urban politics, urbanity is not an imagination of a 'real' urban landscape, but a staged image of urbanity. As a consequence, studies on the urban condition need to recognize not only the *restructuring* of urban space and society, but also the *re-interpretation* of urban space and society.

Economies of Fascination: Strategic Staging of Urban Landscapes as Places and Spaces of Fascination

In order to investigate the increased competition between cities, the mutual relationships between the cultural and the economic spheres have to be reconsidered, both in the sphere of production and consumption: 'Research must grasp the ways in which culture, aesthetics and symbolic processes have interpenetrated with the political and economic' (Jayne 2006: 41). We therefore argue that the analysis of the interplay of film images and imaginations – and of film images and the 'strategic re-cycling' of film images and urban imaginations – offer promising insights.

In looking at the economic and discursive processes underlying the symbolic production and consumption of themed urban environments, we have introduced the relationship between the production and the consumption of signs in films and city marketing as a field where the categories of place and space are intertwined. The city of Hamburg is an instructive example in this sense. Our argument shows that the combination of semiotic and discourse-theoretical approaches analyzing urban images in films and film images in urban development only touch the surface. Whereas the semiotic perspective concentrates on how film images serve as signs – and create meanings as distinctive signifiers – the discursive approach is limited to the effects of these signs, i.e. looking at the ways the images and their meanings are 'used' ('re-cycled') in city-marketing strategies. Here, fascination appears as a form of discursive strategy that obscures seemingly self-evident meanings in the process of signification by employing superficial emotional links. A critical perspective on media geography thus treats 'media not merely as a visual reflection of the material world but as discrete moments in the production and circulation of cultural meaning' (Craine 2007: 149).

Our examples from Hamburg exemplify the fact that in films Hamburg is used in the sense of a 'landscape as place', but the incorporation of these images into the strategic policies of the HafenCity corporation makes it an iconographic signifier that is turned into a 'landscape as space'. On the other hand, local administration's present city-marketing strategies try to turn this space again into place: thus, the images and imagination of the HafenCity are re-cycled in order to serve as signs for a distinctive *local* feature (Hamburg *as place*) – despite its rather *global* architecture.

'The repetitious use of icons by film and television of particular places and buildings can create a representational legacy that works to construct and establish a cognitive map, a sense of place' (Lukinbeal 2005: 8): It thus seems that the postmodern, yet rather interchangeable image of the HafenCity and the 'Elbe string of pearls' is strategically used to 'stage' and to advertise the city of Hamburg. And by doing so, this strategy follows the general model in which Hamburg labels itself as a 'Green Service Metropolis'. But this model is based on both the city's long tradition as a commercial city and port and the fact that its economy is nowadays shaped by services. The waterfront and the port area on the banks of the River Elbe are still distinctive spatial features and the waterfront development projects called *HafenCity* and *String of Pearls* on the waterfront aim to represent the place's tradition. Thus, even though promoting postmodern style and architecture, European cities like Hamburg use the image of tradition and postmodernity together to mark the fascination of European urban landscapes by playing the game of place and space.

References

Aitken, S. and Dixon, D. 2006. 'Imagining Geographies of Film'. *Erdkunde*, 60(4), 326–336.

Aitken, S. and Zonn, L. 1994. *Place, Power, Situation and Spectacle. A Geography of Film*. Lanham: Rowman & Littlefield.

AlSayyad, N. 2006. *Cinematic Urbanism: A History of the Modern from Reel to Real*. London: Routledge.

Benjamin, W. 1980. 'Das Kunstwerk im Zeitalter seiner technischen Reproduzierbarkeit', in *Gesammelte Schriften I*. Frankfurt am Main: Suhrkamp, 471–508.

Barthes, R. 1967. *Elements of Semiology*. London: Cape.

———— 1973. *Mythologies*. London: Granada.

———— 1977. *Image, Music, Text*. London: Fontana.

———— 1982. *Empire of Signs*. London: Jonathan Cape.

Baudrillard, J. 2004. *Amerika*. Berlin: Matthes & Seitz.

Bollhöfer, B. 2007. *Geographien des Fernsehens. Der Kölner Tatort als mediale Verortung kultureller Praktiken*. Bielefeld: Transcript.

Bollhöfer, B. and Strüver, A. 2005. 'Geographische Ermittlungen in der Münsteraner Filmwelt: Der Fall Wilsberg'. *Geographische Revue*, 7(1–2), 25–42.

Clarke, D. (ed.) 1997. *The Cinematic City*. London: Routledge.

Craine, J. 2007. 'The Medium has a New Message: Media and Critical Geography'. *International E-journal for Critical Geographies (ACME)* [Online], 6(2), 147–52. Available at: http://www.acme-journal.org/vol6/JCr.pdf [accessed: 5 July 2010].

Culler, J. 1976. *Saussure*. London: Fontana.

de Certeau, M. 1984. *The Practice of Everyday Life*. Berkeley: University of California Press.

de Saussure, F. 1960. *A Course in General Linguistics*. London: Owen.

Donald, J. 1999. *Imagining the Modern City*. London: Athlone Press.

du Gay, P. and Pryke, M. (eds) 2002. *Cultural Economy*. London: Sage Publications.

Duncan, J. and Ley, D. (eds) 1993. *Place/Culture/Representation*. London: Routledge.

Flitner, M. and Lossau, J. 2005. *Themenorte*. Münster: LIT Verlag.

Ford, L. 1994. 'Sunshine and Shadow: Lighting and Color in the Depiction of Cities in Film', in *Place, Power, Situation and Spectacle. A Geography of Film*, edited by S. Aitken and L. Zonn. Lanham: Rowman & Littlefield, 119–36.

Foucault, M. 1973. *The Order of Things*. New York: Vintage.

———— 1980. *Power/Knowledge*. Brighton: Harvester Press.

Gibson-Graham, J.K. 2000. 'Poststructural Interventions', in *A Companion to Economic Geography*, edited by E. Sheppard and T.J. Barnes. Oxford: Blackwell Publishers, 95–110.

Gold, J. and Ward, S. (eds) 1994. *Place Promotion: The Use of Publicity and Marketing to Sell Towns and Regions*. New York: John Wiley.

Gottdiener, M. 2001. *The Theming of America. Dreams, Media Fantasies, and Themed Environments*. Boulder, Oxford: Westview Press.

Hall, S. 1980. 'Encoding/Decoding', in *Culture, Media, Language*, edited by S. Hall. London: Routledge, 128–38.

———— 1997. *Representation: Cultural Representations and Signifying Practices*. London: Sage Publications.

Harvey, D. 1990. *The Conditions of Postmodernity. An Enquiry into the Origins of Cultural Change*. Cambridge, Oxford: Blackwell Publishers.

Hopkins, J. 1994. A Mapping of Cinematic Places: icons, ideology, and the power of (mis)representation, in *Place, Power, Situation and Spectacle. A Geography of Film*, edited by S. Aitken and L. Zonn. Lanham: Rowman & Littlefield, 47–65.

Jayne, M. 2006. 'Cultural Geography, Consumption and the City'. *Geography*, 91(1), 34–42.

Katz, C. 1999. *Manhattan on Film: Walking Tours of Hollywood's Fabled Front Lot*. New York: Limelight Editions.

Kennedy, C. and Lukinbeal, C. 1997. 'Towards a Holistic Approach to Geographic Research on Film'. *Progress in Human Geography*, 21(1), 33–50.

Krätke, S. 2002. 'Produktionscluster in der Filmwirtschaft. Beispiel Potsdam/ Babelsberg'. *Zeitschrift für Wirtschaftsgeographie*, 46(2), 107–23.

Lash, S. and Urry, J. 1994. *Economies of Sign and Space*. London: Sage Publications.

Lukinbeal, C. 2004. 'The Rise of Regional Film Production Centers in North America, 1984–1997'. *GeoJournal*, 59(4), 307–21.

————— 2005. 'Cinematic Landscapes'. *Journal of Cultural Geography*, 23(1), 3–22.

Lukinbeal, C. and Zimmermann, S. 2006. 'Film Geography. A New Subfield'. *Erdkunde*, 60(4), 315–26.

Lütke, P. 2006. *Kreative Produktionsmilieus in der Film- und Fernsehwirtschaft. Content-Produktion in Köln*. Münster: LIT Verlag.

Mattissek, A. 2008. *Die neoliberale Stadt: Diskursive Repräsentationen im Stadtmarketing deutscher Großstädte*. Bielefeld: Transcript.

Rose, G. 2001. *Visual Methodologies: An Introduction to the Interpretation of Visual Materials*. London: Sage Publications.

Strüver, A. 2005. *Stories of the 'Boring Border': The Dutch-German Borderscape in People's Minds*. Münster: LIT Verlag.

Waitt, G. 2005. 'Doing Discourse Analysis', in *Qualitative Research Methods in Human Geography*, edited by I. Hay. Oxford: Oxford University Press, 163–91.

Ward, S.V. 1998. *Selling Places. The Marketing and Promotion of Towns and Cities*. London: Routledge.

Zimmermann, S. 2003. 'Reisen in den Film – Filmtourismus in Nordafrika', in *Tourismus – Lösung oder Fluch? Die Frage nach der nachhaltigen Entwicklung peripherer Regionen*. Mainzer Kontaktstudium Geographie, vol. 9, edited by H. Egner. Mainz: Department of Geography, University of Mainz, 75–83.

Zukin, S. 1995. *The Cultures of Cities*. Oxford, Cambridge: Blackwell.

Filmography

Berlin: Symphonie einer Großstadt (dir. Walther Ruttmann, 1927).

Metropolis (dir. Fritz Lang, 1927).

Modern Times (dir. Charlie Chaplin, 1936).

Paulas Geheimnis (dir. Gernot Krää, 2006).

Die Entscheidung (dir. Nikolaus Leytner, 2005).

Playtime (dir. Jacques Tati, 1967).

Mon Oncle (dir. Jacques Tati, 1958).

Blade Runner (dir. Ridley Scott, 1982).

Manhattan (dir. Woody Allen, 1979).

Falling Down (dir. Joel Schumacher, 1993).

The Day After Tomorrow (dir. Roland Emmerich, 2004).

The Truman Show (dir. Peter Weir, 1998).

Chapter 11

'Neoliberalism with Chinese Characteristics': Consumer Pedagogy in Macao

Tim Simpson

In 2007, LifeTec Group, a Hong Kong biopharmaceutical company specializing in drugs to treat Hepatitis B and liver fibrosis, diversified into design and production of hybrid gaming machines for the casino industry in the nearby Chinese city of Macao. As a result, the company's stock value tripled and they changed their name to Paradise Entertainment Ltd. The company's initial product was a gaming machine called 'Live Baccarat' that combines an electronic baccarat game with a live dealer. Sitting on a podium in front of a bank of electronically linked individual gaming terminals, the croupier deals a hand of baccarat; the cards are displayed both on a large video screen in the front of the casino and on small screens at the individual terminals where each player sits. As many as 75 players make simultaneous bets via electronic touch screens on their gaming terminals. In December, 2007, the company opened a Paradise Casino in Macao under a profit-sharing agreement with gaming license holder *Sociedade de Jogos de Macau* (SJM), exclusively housing 280 Live Baccarat machines.

Live Baccarat is an innovative attempt to create an electronic casino game that will appeal to Chinese gamblers in Macao. Slot machines and electronic games had previously generated little interest among Chinese gamblers for several reasons: their relatively low payouts when compared to table games; suspicion that such games were rigged in favour of the casinos; and the preference of many Chinese gamblers to play against a live dealer so they can employ rituals or strategies of 'luck' that allow them to feel they can manipulate the outcome of the game (Lam 2006, 2008). For the casino, the appeal of an electronic game like Live Baccarat is the ability of the technology to both speed up play and reduce labour and operating costs. Where a typical baccarat dealer can theoretically deal 500 hands in 24 hours, to a table of 6–9 players, the LIVE Baccarat dealer can deal 1000 hands in 24 hours to 75 players simultaneously. This means that the casino using the LIVE Baccarat system can service up to 20 times more gamblers per day, thereby significantly reducing labour costs. The lower labour costs make it feasible for the casino to lower the minimum bet on a table (HK50 at Live Baccarat compared to HK100 for ordinary baccarat) which expands the casino's ability to attract ordinary 'mass market' gamblers. This is a departure for Macao casinos, which have historically been promoted to big-stakes VIP gamblers.

Nearly 27 million tourists visited Macao in 2007, more than half of them from mainland China. These tourists helped make the tiny city of Macao the most lucrative gambling site in the world. In 2007, casino revenues in Macao surpassed those of Las Vegas and Atlantic City combined. The ability of tourists from the People's Republic of China to travel to Macao has been facilitated by the central government's introduction of the Individual Visitation Scheme (IVS), which allows residents from select provinces of the PRC to travel to the former colonies and Special Administrative Regions (SAR) of Macao and Hong Kong without the sanction of an officially-sponsored tour group. For many Chinese tourists, Macao is the first destination they visit outside the mainland. Others make repeated trips to Macao to gamble in the city's casinos. For these tourists, Macao serves as a *laboratory of consumption* in which a nascent post-socialist *neoliberal Chinese subject* is forged. Electronic games like Live Baccarat created for the Macao market can be understood as technologies of neoliberal subjection.

Neoliberalism is generally understood as a mode of governance indicative of contemporary western liberal democracies, where state intervention in such areas as income redistribution or social welfare programmes recedes to be compensated for by privatization of services within a market-driven framework and increased responsibility placed on individuals to care for and regulate themselves (Harvey 2006). To speak of 'Chinese neoliberalism' may seem oxymoronic given that China is an authoritarian state with a planned economy and even under economic reforms differs markedly from neoliberal regimes like those found in the United States or England. However, it is constructive to understand Chinese tourism to Macao within the rubric of neoliberal governance. We will see that Macao can be understood as a 'space of exception' (Ong 2006: 7) where China exercises 'neoliberal interventions' even though for the nation at large 'neoliberalism itself is not the general characteristic of technologies of governing' (Ong 2006: 3, see also Rofel 2007, Harvey 2006, Sigley 2006). In this space of exception, Chinese visitors have the opportunity to engage in travelling, gambling, and shopping activities that, taken together, create the corporeal 'conditions of possibility' for neoliberalism. Attention to tourism and casino gambling in Macao reveals how a new spatial aesthetic of themed environments provides the context for new forms of neoliberal governance, based on the experience of new technologies of 'freedom' and individuation through the formation of emotional ties and fascination as a form of neoliberal subjectivism. At the same time, these new freedoms are also indicative of complementary forms of control through logics of 'dispossession' (Harvey 2005). Indeed, the experience of casino gambling in Macao both embodies neoliberal subjectivation and produces and directs a neoliberal body.

Spatial Aesthetics: Themed Environments in Macao

The ability of mainland Chinese citizens to gamble in Macao is a product of the recent political transformation of the city. Portugal returned the territory of Macao to the

People's Republic of China in 1999, after nearly 450 years of colonial administration. Following the handover, Macao's post-colonial government dismantled the local gambling monopoly controlled for 40 years by Hong Kong billionaire Stanley Ho, and opened Macao to investment by foreign gaming companies from North America, Australia, and Hong Kong. These companies are collectively spending twenty-five billion dollars to tap the increasingly affluent and mobile market of tourists just across the border in mainland China who can travel to Macao under the IVS.

To understand Macao as a 'laboratory of consumption', we must situate the city within the Pearl River Delta Region (PRD) of China. This region includes nine prefectures of Guangdong province, as well as the Special Administrative Regions (SAR) of Hong Kong and Macao, and is home to more than 40 million people. Adjacent to Macao and Hong Kong, respectively, are the Special Economic Zones (SEZ) of Zhuhai and Shenzhen, which Deng Xiaoping established in 1979 as key components of China's economic reforms. The SEZs were ostensibly established as 'laboratories of production', where China would allow foreign direct investment and joint ventures and would study capitalist management and production techniques, as well as testing a private real estate market (Chuihua et al. 2001). The SEZs and the SARs are enclaves in which China applies what Ong calls 'zoning technologies'.

> 'Zoning Technologies' refers to political plans that rezone the national territory. The technologies of governing are the instrumentalization of a form of market-driven rationality that demarcates spaces, usually nonadjacent to each other, in order to capitalize on specific locational advantages of economic flows, activities, and linkages (Ong 2006: 103).

Today the PRC central government pursues economic growth in part by promoting domestic consumption, and the SEZs and SARs make this project of Chinese consumerism viable. Tourism, shopping, and casino gaming are primary consumption activities. In this sense, Macao can be understood as both a product, and a progenitor, of China's economic reforms. As a 'space of exception', Macao provides one technological link between mainland market-socialism and western capitalism.

The majority of the foreign investment in Macao has come from North American gaming companies Venetian Sands Inc., Wynn Inc., and MGM Grand. Venetian Sands has announced plans to invest approximately twelve billion dollars in Macao, nearly three billion dollars of which was spent constructing the two largest casinos in the world in the city. With the Sands Casino – the initial foreign foray into the territory – the company aimed to tap the huge potential 'mass market' of Chinese gamblers, thereby departing from the former gaming model in Macao that relied almost exclusively on big-spending VIP visitors who gamble in private rooms. In contrast, the Sands Casino was the first 'Las-Vegas style' casino in Macao, built with the large stadium-style open gaming floor typical of casinos in the United States. Other gaming companies followed suit, designing not only large gaming floors but housing them in iconic themed buildings of a type previously absent from the Macao skyline, featuring motifs such as Greek Mythology, Venice, and Babylon (fig. 11.1).

Figure 11.1   **Grand Lisboa Casino and Hotel (right), next to Bank of China building, Macao 2009. The architect notes that the iconic shape of the building is reminiscent of the headdress of a Las Vegas showgirl, while the owner of the property claims it was meant to resemble a lotus, the symbol of Macao.**

Source: Schmid 2009

This new Las Vegas-style approach to gaming in Macao has affected not only the architecture and interior design of casinos but also the games on offer. Although the majority of Macao casino revenues are still generated by VIP gaming – particularly high stakes baccarat – there is a concerted effort by casinos to attract mass market gamblers. Companies engage in promotional activities to enable easy access to gambling facilities that include free bus transportation to casinos or the distribution of cards valid for one free slot machine pull in the casino. Further, foreign companies have significantly increased slot machines and electronic gaming formats in an effort to both draw in new members of the mass market for whom the games are easier and therefore less threatening than table games, and to reduce labour and operating costs. The number of slot machines in Macao rose from 814 in 2000 to 7349 by the end of the first quarter of 2007, with 50,000 slot machines projected for the city by 2010. The Venetian casino alone has 3500 slot machines. In addition to standard slot machines, Macao casinos are promoting electronic game innovations including server-based slot machines, LED-screen animated dealers, and new hybrid game formats such as Live Baccarat that integrate live croupiers with electronic gaming terminals. Thus, these casinos use new gaming technologies to attract an expanded population of gamblers who are embedded in aesthetic environments.

Governmentality: The Production of a Neoliberal Subject

It is easy to dismiss themed casino environments based around iconic representations of Venice or ancient Greece as 'meaningless', given that they have no historical or cultural relation with a Portuguese city in China. Indeed, there is a long line of postmodern semiotic criticism of such hyperreal simulations (c.f. Eco 1986, Baudrillard 1983). However, if we only attend to the spatial and architectural *form* of these new Macao casinos, we overlook their *function* for visitors. Likewise, it is tempting to assume that the introduction of Las Vegas casino design techniques and slot machine games into Macao is a reflection of the inevitable global expansion of American culture realized at its most base and gaudy level – that is, that Macao is 'learning from Las Vegas' (see Venturi, Brown and Izenour 1977). To talk about 'Las Vegas in Asia' is misleading, however, because Macao retains unique characteristics that distinguish it from Las Vegas – particularly in VIP gaming (see Leong 2002) – but more importantly because the role the city plays in China's economic development arguably overshadows the casinos' financial contributions to the fortunes of American entrepreneurs. These new entertainment venues in Macao should not be dismissed as empty simulacra, nor are they part of a project to colonize foreign lands for the purpose of geographic or cultural expansion. Rather, they are indicative of a 'new imperialism' (Harvey 2005) that entails a neoliberal project to open and mould global markets to create new opportunities for investment of excess capital. The introduction of American casinos to East Asia can be understood as one trajectory of this logic.

Post-handover Macao is one site for what Harvey calls the 'spatial fix' for excess capital. Fixing capital in geographic space entails not just opening 'free' markets such as Macao's gaming industry and investing in related infrastructure, but the making of new subjects who have been 'disciplined' to participate in them. These subjects must embody the tenets of neoliberalism.

> In so far as neoliberalism values market exchange as 'an ethic in itself, capable of acting as a guide for all human action, and substituting for all previously held ethical beliefs', it emphasizes the significance of contractual relations in the marketplace. It holds that the social good will be maximized by maximizing the reach and frequency of market transactions, and it seeks to bring all human action into the domain of the market. This requires technologies of information creation and capacities to accumulate, store, transfer, analyse, and use massive databases to guide decisions in the marketplace. Hence neoliberalism's intense interest in and pursuit of information technologies (leading some to proclaim the emergence of a new kind of 'information society') (Harvey 2006: 3f.).

The movements of capital invested in Macao not only result in a new landscape of themed casinos on reclaimed land, but also entail a tertiary didactic element – the 'education' of formerly socialist Chinese subjects to consume themed environments, play electronic games, and behave like bourgeois tourists. Opening China's markets for foreign investment means directing individuals towards a neoliberal form of life which can be realized in the spaces of exception like Macao, if not elsewhere. This inadvertent function of Macao's casinos is one characteristic that distinguishes them from those in Las Vegas. Macao's casinos join other sites and discourses of neoliberal consumer pedagogy in China, including state-sponsored tourism (Nyiri and Breidenbach 2007), consumer oriented television shows (Hua 2007), and reality television programming (Keane, Fung and Moran 2007).

The remainder of this chapter will sketch out in detail what Harvey (2006) provocatively calls 'neoliberalism with Chinese characteristics' in the example of casino gambling in Macao. That is, I aim to bridge the sort of Marxian political economic analysis of neoliberalism pursued by Harvey with a Foucaultian account of subjection, or subject formation. It is an attempt to answer 'the question of how top down initiatives to neo-liberalize economies, institutions, or spaces actually work out in practice' (Barnett et al. 2008: 625).

The Phenomenology and Affect of Gambling in Macao

Understanding the comportment of a neoliberal Chinese subject requires attending to the experiential dimension of tourism and casino gambling. The argument proceeds by linking increased freedoms in China with Chinese gambling behaviour and new electronic games recently introduced in Macao's casinos. Taken together, these regulatory discourses and corresponding behaviours constitute the 'conditions of possibility' of neoliberal subjectivation. They do so by promoting different dimensions of fascination, closely intertwining structural-functional conditions with individual emotional and socio-psychological responses. As detailed below, this process includes: attracting consumers, particularly mainland Chinese, by new 'liberties' of mobility; the creation of a socio-cultural environment for leisure-labour activities that permits highly risky speculative performances; and the parallel organization of structural financial transactions and subject-centred behaviour that embeds the individual in a network of debts and obligations. All of these processes come together in the context of new casino technologies that situate the consumer's body in a human-machine interface for the purpose of privatized bodily comfort which ultimately facilitates a more rapid rate of gambling.

Mobility and Individualism

Cross-border Chinese tourist mobility via the Individual Visitation Scheme is a corporeal exercise of freedom of movement, to the extent that a trip to gamble and shop in Macao is ironically experienced as the expression of individual 'freedom'. Of course, one must be cautious when speaking of individual liberties in China, even under a market-socialist economy. China is far from a liberal democracy, and those freedoms that are granted are typically apportioned out by party planners in a careful and methodical manner. Indeed, this is what makes them amenable to study. With the legacy of a work unit (*danwei*) system and household registration (*hukou*) that segregates urban and rural residents, freedom of movement within the nation – not to mention across borders – has been tightly regulated and restricted in the PRC. It is against this background of constraints on mobility that the IVS becomes particularly interesting. Under this scheme, which was introduced in July 2003, many Chinese citizens were for the first time granted permission to leave China proper and travel to the SARs of Macao and Hong Kong. Therefore, the tourist trip to Macao becomes the exercise of a newly-granted individual liberty made possible for some Chinese by the rise in per capita income created by growth produced by economic reforms. Macao is the only place in China where gambling is legal, thereby combining increased freedom of movement, the relatively 'free' markets of post-handover Macao and reform-era China, and the freedom to gamble.

This freedom of movement has been carefully regulated by the central government. Initially it was granted only to residents of Beijing, Shanghai, and eight cities of Guangdong province, before being gradually extended to the remaining cities of Guangdong as well as nine other cities in Jiangsu, Zhejiang, and Fujian provinces. When the plan was initially introduced, throngs of Chinese tourists descended on Macao. The gaming companies that initially benefitted the most were those that provided free bus transportation from the Macao-Chinese border gate – and even from interior provinces of China – directly to the casino. One such venue was the Greek Mythology Hotel and Casino. During the opening week of operations the Greek Mythology Casino claimed crowds of 30,000 visitors. Most of the tourists milled around the driveway outside the building and took photos of the decorative fountain in front, which features a large plaster Poseidon surrounded by a coterie of bathing beauties and wide-eyed fish. General tourist arrival numbers to Macao have grown significantly year-by-year, from 10 million in 2005, to 17 million in 2006, to 27 million in 2007. When concern grew about the number of tourists entering Macao, the central government made additional alterations to the IVS policy, limiting the number of trips tourists could make per month, the length of time one could stay in Macao while on layover prior to travel to another location, and the ability to travel to both Hong Kong and Macao on the same travel visa. Each change in the policy has immediate repercussions for casino revenues. The IVS is thus a technology of mobility with which the central government is experimenting. We might see this as a distinct form of 'Chinese governmentality', 'which governs not through familiar tactics of "freedom and liberty" but through a distinct planning and administrative rationality' (Sigley 2006: 490).

Mobility is of crucial concern to the PRC; rural to urban migration in China today constitutes the largest human migration in world history. One consequence of increased mobility is destabilization of traditional communities based on proximity and the dis-embedding of individuals from groups. As people move away from local communities they inevitably experience increased individualism. Therefore, freedom of movement also contributes to an enhanced sense of individualism, not in terms of ideology but of behaviour and experience.

Gambling as Labour

Casinos operate like factories that produce no tangible product but the manipulation of consumer affect, principally directed towards excitement and fascination. The excitement of casino gaming obscures the productive pedagogical relevance of the activity for gamblers. For many Chinese players, casino gaming is approached as a mode of *labour* rather than leisure; the primary aim is to make money rather than to have fun (see Lam 2008). In this way, tourists from China visiting Macao's themed casino environments 'work' at being consumers, thus confusing the traditional western social relations of labour and leisure. For example, former stockbroker, Hong Kong legislator, and avid Macao gambler Cham Pui-ching says 'people should not take gambling as entertainment, because they will lose a lot of money if they go to the casino for pleasure. I take gambling as a serious investment and I measure how much I can lose' (Yiu 2008). This feature of Chinese gamblers is often noted in the gaming-oriented journalism in Macao (e.g. *Macau Business, Inside Asian Gaming*). Macao's new entrepreneurs and managers are cautioned to recognize this characteristic of the typical Chinese gambler. In consequence, unlike their Las Vegas counterparts, Macao casinos do not provide gamblers with free alcoholic drinks. Chinese players tend to feel that alcohol will harm their concentration, so casinos in Macao distribute free tea, orange juice, and milk. Thus, the ambiance of Macao casinos is not so much party conviviality as intensity.

Baccarat is by far the most popular game in Macao's casinos and generates the most revenue. It is popular largely because it offers the highest potential return for gamblers. Also, it is a game the outcome of which many Chinese players feel can be manipulated by their intervention. Such players keep careful records of cards dealt in an effort to discover patterns that will allow them to predict future outcomes. In Macao baccarat games, the hold cards are dealt face down (although the game itself does not require this); players 'squeeze' their cards, folding up the corners to peek at the card in order to glimpse whether the face value appears to be high or low, and then engage in behaviours (e.g. banging on the table or chanting) they hope will induce a change in the face value of the card. In this way, the cards are destroyed each hand, like the raw material in a factory. Macao gamblers manage to 'consume' a product in what is generally an affect-exclusive experience.

Gambling as Speculation

Casino gambling can also be understood as an extraordinary ersatz model of capitalist speculation and investment. 'If you want to gamble be more serious and have a strategy like making an investment', Chim advises. 'I find gambling and investing have some similarities' (Yiu 2008). Like the stock market, gambling allows for economic return to be divorced from individual labour. 'The gambler as financial speculator does not resist capitalist prescriptions, but rather fundamentally embodies them', writes Gilloch (1997). 'Capitalism formalizes the activity of the gambler and his desire to make money through the institution of the stock market' (Gilloch 1997: 158).

This seems particularly true for Chinese gamblers whose conceptualization of risk may differ from that of some other cultural groups. Psychologists of risk have differentiated between two types of risk taking behaviour. In situations involving *Excitatory Value* (e.g. gambling) people exhibit Stimulating Risk Taking (SRT). In situations involving *Instrumental Economic Activity* (e.g. investing, education) people exhibit Instrumental Risk Taking (IRT). In one study of European university students (Zaleskiewicz 2001),[1] risk-taking was found to be *situation-specific*. High risk activity in one domain (gambling) was not correlated to the other domain (investment). That is, subjects who were likely to make risky bets were not likely to make risky investments. Studies conducted with mainland Chinese gamblers in Macao, however, find that high risk SRT and IRT *are* significantly related for Chinese (Vong 2007, Ozorio and Fong 2004). Ozorio and Fong (2004: 35) claim that 'instant or quick rewards provide a reason for the Chinese to take risks in gambling and investment'. Chinese gamblers with a propensity for risk are also likely to carry that propensity to stock market investing, and vice versa.

---

1 Zaleskiewicz (2001) does not specify the ethnicity of the subjects in this research. However, he indicates they were undergraduate business students in two psychology courses. Since his affiliation is with the Institute of Organization and Management, Wroclaw University of Technology, Poland and the results are published in a European journal, presumably the students were enrolled in a university in Poland.

Electronic Gambling as Epistemic Consumption Object

The innovation of new networked multi-player game formats and styles of play in Macao mean that increasingly a round of gambling is not a solitary wager but becomes part of a project that is connected to future activity. Electronic games – through such features as 'progressive jackpots' and club benefits for frequent players – constitute an 'epistemic consumption object' (Zwick and Dholakia 2006) that draws gamblers into a project of continuous knowledge activity. The epistemic consumption object is one that features 'material elusiveness' or lacks 'ontological stability', such that the object is not simply consumed at once in its entirety but is transformed into a 'continuous knowledge project' for the consumer (Zwick and Dholakia 2006: 21).

For example, 'progressive jackpots' refer to an optional side bet players may make on some linked electronic games that places the wagered money into a collective pool; when a specific card sequence is dealt on a player's electronic baccarat game, for example, if the player has made the side jackpot bet, then he or she wins the collective jackpot. Macao casinos offer the world's first baccarat-based progressive jackpot. The jackpot may accumulate for weeks or more before someone wins, and therefore the amount of money can be quite significant, enough for it to be called, within the industry, a 'life changing jackpot'. Players who make this side bet are motivated to return to the game system because of the long-term 'progressive' duration of the bet. They have contributed money to the jackpot that can only be won in future games from that set of networked games. Each time the player returns and makes the side bet, the potential jackpot grows larger. Therefore, electronic gambling takes on the form of an investment.

Another example of gambling as an epistemic consumption object can be found in Macao's Mocha Clubs, which are the first local casinos offering solely slot machines and electronic games. Mocha is the brainchild of one of Stanley Ho's sons, Lawrence Ho, who currently operates Melco, a joint partnership between SJM and Public and Broadcasting, Ltd., of Australia. Mocha markets its clubs to younger gamblers who may prefer the cafe-style ambience and whose propensity to play video games may make them more amenable to electronic gaming. In an effort to convince Chinese gamblers to play slot machines, Mocha Clubs produced electronic versions of table games which are essentially a hybrid of table games and slot machines. Many of these electronic games are linked to offer progressive jackpots. Further, Mocha acts as a membership club with four levels of membership. Members are identified by a player's card which they can use for cashless gaming transactions and which accumulates data regarding the number of visits and volume of play. Higher volume allows players to ascend to higher levels of membership with additional benefits. The information can also be used to market special promotions to players based on their game preferences and betting patterns. The ontologically unstable nature of such an enterprise means that many players approach electronic gaming at Mocha as a 'continuous knowledge project'. According to Mocha's Chief Operations Officer, 'Our players are loyal, regular players. They are not one-time players … so they are very smart. When they play several times, they know the maths exactly. And they know which games are their type' (*Moving With Mocha* 2008: 9f.).

A player's focus on such epistemic gaming objects contributes to a heightened individualism. In electronic gaming of this sort, 'consumers experience a type of *sociality with consumption objects*' (Zwick and Dholakia 2006: 20), rather than with other people. Such games increase individualism among players by focusing their attention on an individualized consumption project.

Structural Homology Among Electronic Gaming and Electronic Financialization

There is a structural similarity between new forms of electronic gaming and new forms of electronic financialization emerging in China. Harvey notes that such electronic financialization is a hallmark of a neoliberal economy. Investment bankers and corporate financial officers need to be able to analyze large amounts of financial data and move capital instantaneously in electronic form in order to exploit opportunities. The ordinary person participates in this phenomenon at the mundane level of bank automatic teller machines (ATM), credit and debit cards, internet banking, and internet stock trade. Electronic gaming in Macao should be considered within the context of the relatively recent exposure of Chinese consumers on the mainland to these rudimentary financial technologies. Mainland Chinese banks first installed automated teller machines only about 20 years ago; their number is rapidly increasing today, from 102,000 in 2006, to a projected 233,000 by 2010 (Perez 2007). In addition to standard ATMs, China stands to experience large growth in innovative technologies like biometric ATMs.

> China Construction Bank had installed terminals that allow balance checking, security transactions and sale of lottery tickets, and supported Chinese characters. Some ATMs can transfer money to mobile phones, check several transaction details, allow foreign currency withdrawals and pay mobile phone, utility and television fees (Perez 2007).

Cash machines in casinos serve not only as ATMs, but may also change large bank notes, cash slot vouchers, exchange currency, and make credit card transactions. Machines operated by companies like Global Cash Assess – which plans to make Macao its operations centre – allow players who have exceeded their daily withdrawal limit on an ATM account to borrow money against future account activity (Azevdeo 2007). Playing an electronic table game or slot machine is homologous to forms of electronic banking.

Buying on credit and living with debt is a crucial feature of any consumer-driven economy, and promotion of such an economy entails not only familiarizing people with electronic finance, but also making people comfortable with debt. In the post-Depression United States, for example, macroeconomic goals to encourage domestic consumption were supported by government and banking policies to facilitate home mortgage loans and television programme plot lines that often centred around formerly frugal immigrant families learning that it was acceptable, and even desirable, to buy on credit or to replace home-made domestic products with appliances bought in shops (see Lipsitz 1990). Similarly, in an effort to encourage domestic consumption, the PRC hopes to make Chinese subjects more comfortable with discretionary consumption and deficit spending. On average, Chinese save a larger percentage of their income than citizens of most other nations, a total of more than 40 per cent of GDP (Walker and Buck 2007: 54). But even in the case of debit cards which only allow consumers to spend money they already possess, the move towards an electronic transaction is a step away from a cash-based economy. The casino player's card is essentially a debit card. Players who use electronic gaming machines are practicing behaviours homologous to other forms of electronic finance, familiarizing themselves with movements of electronic capital in the process.

Therefore, travelling to Macao and gambling in a casino puts into practice fundamental neoliberal characteristics – 'free' movements of people and capital, increased individualism, market speculation, consumerism, information technology. In this way, Chinese tourism in Macao can be understood in part as an act of neoliberal subjection that molds a consuming body that can be directed towards other homologous consumption activities. The themed casino functions as a 'normalizing apparatus' (Barnett 2001: 11) that articulates all of these practices with the emerging subject.

Comportment of a Neoliberal Subject

All of these characteristics can be observed in game innovations designed for the Macao market such as Live Baccarat that combine live dealers with electronic gaming terminals. Such game design is motivated by a desire to encourage Chinese gamblers to play slot machines and other self-contained games – which have low operating costs – by familiarizing these gamblers with electronic games that maintain the draw of a face-to-face interaction with a human dealer.

In addition to Live Baccarat, Paradise Entertainment's newest gaming product is the Paradise Box. This version of the game includes the Live Baccarat system of live dealer and linked electronic terminals but features a 'progressive jackpot' betting option, an additional integrated server-based slot machine at each terminal, and an ergonomic seat. The progressive jackpot transforms the game into an epistemic consumption object, while the slot machine allows the player to play slot games between hands of baccarat. A company executive notes that 'It usually takes people ten seconds, maximum, to place a bet, so they have 40 seconds of downtime, and during that downtime they are fidgety.' Therefore, opportunities to make additional electronic bets are integrated into the system between each hand dealt by the live croupier. The executive reports that 'A lot of customers have been complaining there is too much downtime. They say "Hurry up. You guys are supposed to be an electronic table game. Go faster"' (*Paradise Found* 2007: 19).

This sense of urgency is compounded by the feeling Chinese gamblers have that they need to ride a streak before their luck 'runs out' (*Paradise Found* 2007: 19). The ergonomic chair attached to the Paradise Box is designed to be an 'extension of the gambler', and the company likens it to the captain's chair from *The Matrix*, which provided a human-machine interface capable of altering future reality. In this way, we can observe that the production of a Chinese consumer is not only the production of a desiring subject but involves the *dressage* (Lefebvre 2004) or corporeal molding of the consuming body. Further, the ergonomic design of a comfortable gaming chair can also be understood as contributing to the production of an increasingly individualized subject. Richard Sennett (1994) has drawn attention to the relationship between comfort and individualism. In a discussion of the interactions between the body and the city in the historical production of individualism in the West, Sennett (1994) has noted how increasing bodily comfort of chairs and seats in domestic spaces, train carriages, and cafes was implicated in the construction of individualism. 'Comfortable ways to travel, like comfortable furniture and places to rest, began as aids for recovery from the bodily abuses marked by the sensations of fatigue', writes Sennett (1994: 339). Leisure comfort compensated for the pain of the body at labour. 'From its very origins, though, comfort had another trajectory, as comfort became synonymous with *individual* comfort. If comfort lowered a person's level of stimulation and receptivity, it could serve the person at rest in withdrawing from other people' (Sennett 1994: 339).

With the Paradise Box, the combination of an electronic consumption object and an ergonomic chair draws gamblers into an increasingly privatized activity focused on a speculative individual epistemic financial project; at the same time, players sit publicly in a geometrically-arranged group of other gamblers. Gambling in the Paradise Box, then, combines at bodily level the macro-components of the neoliberal economy. This machine demonstrates how, 'In projects of political subjectification or governmental self-formation, appropriate bodily comportments and forms of subjectivity are to be fostered through the positive catalytic qualities of spaces, places, and environments' (Huxley 2007: 195). If we attend to the body which emerges as a product of people's interactions with the built environment, from the themed casino itself to the chair attached to an electronic casino game, we can understand Macao as a *crucible* for the 'alchemical' transformation of formerly socialist subjects into post-socialist, neoliberal consumers. This turns attention away from the 'aesthetics' of the themed casino and the 'desire' for affective pleasures, and to the *kinesthetics* of the body engaged in the *work* of gambling.

The 'Problem' of Gambling: Chinese Gambling as Accumulation by Dispossession

However, one should always understand such microeconomic behaviour within a larger macro-structural context. This becomes particularly evident when we turn to discussions of problem gambling. When academics speak of gambling as a 'problem', the focus is generally on individuals defined as 'problem gamblers' and conceptualized in psychological (see Lin, Raylu and Tian 2007) or neuro-biological terms (Vrecko 2008). This is consistent with neoliberal ideology that seeks to make the individual the locus of decision-making and responsibility. As Nicoll (2007: 116) says, 'the problem gambler is an abstraction of neoliberal governance detached from embodied and specifically located subjects and groups'.

If we look at the 'problem' of gambling sociologically and economically, rather than as an individual pathology or product of brain chemistry, we might understand the attempt to encourage casino gaming consumption activities among mainland Chinese 'mass market' tourists as an instance of what Harvey calls 'accumulation by dispossession' (2005: 144); that is, accumulation of capital through the appropriation, co-optation, or exploitation of public, common resources or the personal property or social or cultural achievements of others. 'Any social formation or territory that is brought, or inserts itself, into the logic of capitalist development must undergo wide-ranging structural, institutional, and legal changes of the sort that Marx describes under the rubric of primitive accumulation', says Harvey (2005: 153). This primitive or original accumulation must then be repeated regularly in order to perpetuate capital accumulation. Harvey (2005: 153f.) notes that 'the turn towards state-orchestrated capitalism in China has entailed wave after wave of primitive accumulation' – unpaid labour, privatization of public land, and so on. The Chinese gamblers' pedagogic 'training' in consumption can be understood as another example; it benefits central bank macroeconomic goals and corporate investors to the detriment of the individual citizen who is exiting socialism to participate in the market. Indeed, individual liberties parcelled out to the subject become the means by which other liberties are taken away. There are at least four ways that this form of gambling exemplifies accumulation by dispossession. These are admittedly not akin to more brutal forms of primitive accumulation, like slavery or theft of natural resources through colonialism, but they still illustrate the inevitable tendencies of this practice.

The first type is illustrated in the comment by a casino executive in Macao that 'the gaming spend in China is not a discretionary spend' (Gough 2008). In other words, the relative amount Chinese tourists spend on gaming in Macao does not fluctuate according to one's means, judgment, or economic ability to absorb the loss. In fact, to say that it is not 'discretionary' is to imply that it is not a choice at all. The casino president contrasts Macao's casino revenues with those of Las Vegas during the current global economic downturn.

> If you look at it you've got the worst winter on record, you've got floods, the earthquake, government policy, a tumbling stock market – and you've got 50-plus per cent growth in Macau gaming revenue. So you've got the macroeconomic and geopolitical environment counted against you and still you have that amount of growth (Gough 2008).

Many Chinese tourists treat casino gaming activity like a necessary, non-discretionary investment in future returns (the average of which are of course inevitably negative due to the mathematic casino house advantage) rather than as a leisure activity. Indeed, the articulation of Macao casinos with the Chinese tendency to approach gaming as a form of work creates a radical form of the 'commodification of labour power' (Harvey 2005: 145), appropriating the productive activity of labour for the purpose of an empty form of consumption in which no product exchanges hands, or an empty mode of financial speculation in which a negative return is inevitable. 'The Chinese are more serious and hard-core gamblers than their western counterparts', contends Vong (2007), 'because when they play, excitement does not seem to be the only reason. Very often, they are propelled by the motive to make money and also perhaps a belief that they can really make money from gambling' (Vong 2007: 40).

A second instance of accumulation by dispossession involves the cooptation of the tertiary educational system in Macao for the benefit of gaming and hospitality industries. Arguably, the top programme (in terms of student interest and admissions qualifications) at the University of Macau, the leading tertiary institution in the city, is the Gaming Management programme. The University of Macau also offers professional certificate programmes for casino staff and has

Figure 11.2   Mock hotel room, Casino Training School, Taipa, Macao, 2007
Source: Lampton 2007

entered into a joint venture with Venetian Sands to teach classes at university facilities on site at the casino property. There are also schools, like the Institute for Tourism Studies and the Casino Training School, which are ostensibly designed to produce staff for service industries. The casino training school not only teaches croupier skills but the proper way for hotel housekeeping staff to make a hotel bed and clean a bathroom (fig. 11.2). This is the flip side of consumer pedagogy; the didactic training of service workers to staff hotels and casinos.

Related to this dispossession of the educational system is dispossession of future life experience for Macao youth. This results from the recent rise in pay for croupiers. Since Macao law dictates that only local Macao residents can work as frontline dealers in the casinos, there is a limited supply of workers to fill entry-level positions. The growth in the industry has led to heightened competition between companies to hire such staff. The result is a significant rise in starting pay, so that a secondary-school graduate can begin work as a croupier, earning a starting salary that is double the Macao median salary (Fox 2008). This attractive alternative to university education has led many youth directly to the gaming industry. The short term gain of a relatively high salary must be weighed, however, against the long-term benefit of higher education, without which there is little chance of career advancement in the local gaming industry above the basic croupier level. Ironically, while foreigners cannot work as croupiers, they may work at the management level on the gaming floor as well as in the organization, so it is unlikely that local youth with secondary school education or less can compete for promotion with educated and experienced staff recruited from abroad. Indeed, there is no motivation for casinos to promote local staff because of the lack of labour to replace them at croupier positions. In this sense, many youth leverage their future for the short-term benefit of the casinos.

Finally, and ironically, we can point to the dispossession of wages via casino gaming. Macao casino workers on average gamble with greater frequency than the rest of the local population. A recent survey revealed that 42 per cent of the casino employees in Macao gamble regularly; in comparison, 23 per cent of the general population of Macao are regular gamblers (Leong 2007). In this way, once in the industry many employees essentially return a portion of their salary to the casino companies and indirectly to the government in the form of gaming revenue taxes. This is especially a problem for the younger staff, who work odd hours due to the 24 hour operation of the casinos and who find their leisure time often comes late at night when there are few opportunities for public socializing with peers outside casinos.

Recent development of themed casinos and electronic game innovations in Macao contribute not only to a phantasmagoric city of fascination, desire, and excitement. These casinos are a product of 'neoliberal interventions' into China's economy. In turn, these economic reforms contribute to production of a neoliberal Chinese subject who is increasingly encouraged to engage in consumption activities and who is educated to do so. As detailed above, promoting state economic growth by encouraging consumption via pedagogical training in consumerism amounts to accumulation by dispossession; it appropriates economic resources gained by individuals as a by-product of the very economic reforms that enabled the growth in the first place. This exemplifies the neoliberal tendency 'to bring all human action into the domain of the market' (Harvey 2006: 4). PRC policies that increase individual freedoms (of inter- and cross-border mobility, for example) are motivated by their benefit to macroeconomic goals of the central government and the profit margins of casinos (and government tax revenues) in Macao. To wed freedom to travel and gamble with the strategies of dispossession reveals the dialectic of subjection: 'the hinge between neoliberalism as exception and exception to neoliberalism, the interplay among technologies of governing and of disciplining, of inclusion and exclusion, of giving value or denying value to human conduct' (Ong 2006: 5). The emergent Chinese subject is herself a self-governing technology of capital accumulation, both the expression of new corporeal freedoms and the implication of the individual body in new forms of power and control. In this case, the fascinating qualities of the city of Macao obscure its function for China's economy.

References

Azevdeo, P. 2007. 'Cashing In'. *Macau Business*, 37(5), 58–61.

Barnett, C. 2001. 'Culture, Geography, and the Arts of Government'. *Environment and Planning D: Society and Space*, 19(1), 7–24.

Barnett, C., Cloke, P., Clarke, N. and Malpass, A. 2008. 'The Elusive Subjects of Neo-liberalism: Beyond the Analytics of Governmentality'. *Cultural Studies*, 22(5), 624–53.

Baudrillard, J. 1983. *Simulations*. New York: Semiotexte.

Chuihua, J.C., Inaba, J., Koolhaas, R. and Sze T.L. (eds) 2001. *Great Leap Forward*. Cambridge: Harvard Design School.

Eco, U. 1986. *Travels in Hyperreality*. Orlando: Harcourt Brace & Co.

Fox, Y.H. 2008. 'Casino Layoffs Give Teachers Chance to Push Education'. *South China Morning Post*, 4 July, C–5.

Gilloch, G. 1997. *Myth and Metropolis: Walter Benjamin and the City*. London: Polity Press.

Gough, N. 2008. 'SJM Boss Has the Skills to Deliver Counterpunch'. *South China Morning Post*, 15 September, B–12.

Harvey, D. 2005. *The New Imperialism*. Oxford: Oxford University Press.

——— 2006. *A Brief History of Neoliberalism*. Oxford: Oxford University Press.

Hua, X. 2007. 'Brand New Lifestyle: Consumer-oriented Programmes on Chinese Television'. *Media, Culture & Society*, 29(3), 363–376.

Huxley, M. 2007. 'Geographies of Governmentality', in *Space, Knowledge and Power: Foucault and Geography*, edited by J.W. Crampton and S. Elden. Aldershot: Ashgate, 185–204.

Keane, M., Fung, A.Y.H. and Moran, A. 2007. *New Television, Globalisation, and the East Asian Cultural Imagination*. Hong Kong: Hong Kong University Press.

Lam, D. 2006. 'Slot or Table? A Chinese Perspective'. *UNLV Gaming Research & Review Journal*, 9(2), 69–72.

———— 2008. 'An Observation Study of Chinese Baccarat Players'. *UNLV Gaming Research & Review Journal*, 11(2), 63–73.

Lefebvre, H. 2004. *Rhythmanalysis: Space, Time and Everyday Life*. Translated from French by S. Elden and G. Moore. London: Continuum.

Leong, A.V.M. 2002. 'The "Bate-Ficha" Business and Triads in Macau Casinos'. *QUT Law and Justice Journal*, 2(1), 83–96.

Leong, S. 2007. 'Survey Finds 42 pct. of Casino Workers Gamble'. *Macau Post*, 23 May, 1.

Lin, J., Raylu, N. and Tian, P.S.O. 2007. 'Validation of the Chinese Version of the Gambling Urges Scale (GUS–C)'. *International Gambling Studies*, 7(1), 101–11.

Lipsitz, G. 1990. *Time Passages: Collective Memory and American Popular Culture*. Minneapolis: University of Minnesota Press.

Moving With Mocha 2008. *Inside Asian Gaming*, 2(3), 6–12.

Nicoll, F. 2007. 'The Problematic Joys of Gambling: Subjects in a State'. *New Formations*, 63(2), 103–20.

Nyiri, P. and Breidenbach, J. 2007. '"Our Common Heritage": New Tourist Nations, Post-socialist Pedagogy, and the Globalization of Nature'. *Current Anthropology*, 45(2), 322–30.

Ong, A. 2006. *Neoliberalism as Exception: Mutations in Citizenship and Sovereignty*. Durham: Duke University Press.

Ozorio, B. and Fong, D.K. 2004. 'Chinese Casino Gambling Behaviors: Risk Taking in Casinos vs. Investments'. *UNLV Gaming Research and Review Journal*, 8(2), 27–38.

Paradise Found 2007. *Inside Asian Gaming*, 1(6), 18–21.

Perez, B. 2007. 'Mainland ATM Usage Forecast to Rise Rapidly'. *South China Morning Post*, 3 December, B–5.

Rofel, L. 2007. *Desiring China: Experiments in Neoliberalism, Sexuality, and Public Culture*. Durham: Duke University Press.

Sennett, R. 1994. *Flesh and Stone: The Body and the City in Western Civilization*. New York: W.W. Norton.

Sigley, G. 2006. 'Chinese Governmentalities: Government, Governance and the Socialist Market Economy'. *Economy and Society*, 35(4), 487–508.

Venturi, R., Brown, D.S. and Izenour, S. 1977. *Learning from Las Vegas: The Forgotten Symbolism of Architectural Form*. Cambridge, MA: MIT Press.

Vong, F. 2007. 'The Psychology of Risk-taking in Gambling among Chinese Visitors to Macau'. *International Gambling Studies*, 7(1), 29–42.

Vrecko, S. 2008. 'Capital Ventures into Biology: Biosocial Dynamics in the Industry and Science of Gambling'. *Economy and Society*, 37(1), 50–67.

Walker, R. and Buck, D. 2007. 'The Chinese Road: Cities in the Transition to Capitalism'. *New Left Review*, 46(4), 39–66.

Yiu, E. 2008. 'Lawmaker Takes the Sting Out of Chance'. *South China Morning Post*, 5 October, 15.

Zaleskiewicz, T. 2001. 'Beyond Risk Seeking and Risk Aversion: Personality and the Dual Nature of Economic Risk Taking'. *European Journal of Personality*, 15(S1), 105–22.

Zwick, D. and Dholakia, N. 2006. 'The Epistemic Consumption Object and Postsocial Consumption: Expanding Consumer–Object Theory in Consumer Research'. *Consumption, Markets and Culture*, 9(1), 17–43.

PART IV
Consequences of Fascination:
New Horizons

Chapter 12
Excess, Fascination and Climates

John Urry

Introduction

This chapter examines some major social changes relating to contemporary conditions of life upon earth. The proliferation of new forms of production is transforming global capitalism, from a capitalism of production to a capitalism of fascination. Under neo-liberalism, the production sphere appears to be no longer constrained by its resource base but relies on the pro-active production of an endless series of new desires in new places. These seem to be producing consumers subjected to the multiple economies of fascination and, on occasions, excess.

In particular, the chapter deals especially with emerging social contradictions that stem from such shifts within contemporary capitalism, from societies of discipline to societies of control, from specialized and differentiated zones of consumption to mobile, de-differentiated consumptions of fascination and excess, and more generally from low carbon to high carbon societies. In focusing upon emergent contradictions I suggest that capitalism, the only game in town, is its own 'gravedigger' as Marx and Engels famously argued (1888 [1848]). But the 'gravedigging' is not being effected by the proletariat becoming 'concentrated in greater masses' as Marx and Engels projected in their scenario building (1888 [1848]: 64). Rather the gravedigging is being brought about by multiple mobilities, 'excessive' global consumption and rising carbon emissions that seem to be destroying the *global* conditions of life upon earth. Marx and Engels wrote how modern bourgeois society: 'is like the sorcerer, who is no longer able to control the powers of the nether world whom he has called up by his spells' (1888 [1848]: 58). I examine below how, through major emergent contradictions, contemporary capitalism is bringing through climate change: 'disorder into the whole of bourgeois society, endanger[ing] the existence of bourgeois property' (Marx and Engels 1888 [1848]: 59).

I consider first how Marx is to be understood as an analyst of complex non-equilibrium systems and draw some implications for economic and social change. I then examine the significance of positive feedback and nonlinearity within the sciences and social sciences of climate change. In the next section I examine the emergent nature of societies of consumption, fascination and excess. I consider some climate change consequences and show in conclusion how capitalism is its own gravedigger and how difficult it is going to be to reverse systems moving 'the world' in seemingly inexorable fashion towards a climate change abyss.

On Contradictions

Marx's analysis of the unfolding 'contradictions' of capitalist production is an early example of complexity analysis within the social sciences. As biologist John Maynard Smith states: 'I think the reason why they [Marx and Engels] were dialectical materialists was that they were trying to understand ... complex systems in a world in which there was no mathematical language ... that they could use to describe them' (1994: 688f.). They argued how the: 'need for a constantly changing market chases the bourgeoisie over the whole surface of the globe. It must nestle everywhere ... establish connexions everywhere' (Marx and Engels 1888 [1848]: 54). This putative globalization resulted in the transformation from feudal and mercantile production towards a more quantitative money-based system. Capitalist enterprises maximized their profits through paying their workers as little as feasible, or making them work as long as possible, or replacing their labour power with capital. Iterated actions reproduced the capitalist system and its emergent class relations and substantial profits were generated, so offsetting the tendency of the rate of profit to fall. This 'exploitation' continues unless collective action by trade unions or the state prevents it. Each capitalist firm operates under these non-equilibrium conditions and responds to 'local' sources of information and opportunity (Cilliers 1998: 5). The emergent contradictions result from iterations taking place at a 'local' level but their effects do not remain local, and produce systemic contradictions.

Thus Marx brings out how there is no tendency for systems to move towards equilibrium. Indeed the very adaptations and co-evolutions that involve the making of profit are simultaneously those that generate capitalist crises and the strengthening of a proletariat. Metaphorically this can be characterized as 'the genie is let out of the bottle and cannot be returned.' Thus, equilibrium models found in most economic system analyses are flawed, especially so-called general equilibrium models (Beinhocker 2006: Chapter 2, 3). Thus no distinction should be made between states of equilibrium and states of growth – all systems should be viewed as dynamic, processual and involving systemic contradiction (Beinhocker 2006: 66f.). And in this chapter I argue that the high carbon economy/society of the twentieth century, focused upon the creation of new needs, new desires and fascinating environments especially during the neo-liberal period at the century's end, is a new emergent contradiction. As such, the contradiction that twentieth century capitalism generated cannot be 'easily put back into the bottle' since it seems complicit within what had been presumed highly unlikely, actually to be involved in changing climates.

On Climates

These arguments about economic non-equilibrium systems and contradictions relate to the most significant issue which current societies are facing, namely global climate change (see the *Theory, Culture and Society* special issue on *Changing Climates*, edited by Szerszynski and Urry (2010)). Here the main issues are first, how human practices have given rise to increased carbon emissions, and second, how those practices may be 'modified' so as to reduce future carbon emissions. Notions of complex systems, feedbacks and nonlinearities are central to various scenarios of future climate change (IPCC 2007, Pearce 2007). Rial et al. summarize how the earth's climate system is: 'highly nonlinear: inputs and outputs are not proportional, change is often episodic and abrupt, rather than slow and gradual, and multiple equilibria are the norm' (2004: 11).

Although these nonlinearities are significant there has been some reduction in the uncertainties involved in the sciences of climate change (see http://www.ipcc.ch/). While the scale and impact of future temperature changes are much debated and, especially in the US and certain developing societies, contested, there is a growing consensus as to the range of outcomes that stem from already released carbon emissions. World temperatures have risen by at least 0.74° C over the past century and this is the product of many forms of human practice that have raised the levels of greenhouse gas emissions in the atmosphere (Stern 2006: ii). Greenhouse gas levels and world temperatures will further significantly increase over the next few decades; and these rises will raise temperatures, some believe by as much as six degrees centigrade (Lynas 2007). In part this will result from multiple forms of positive feedback rather similar to the kind of non-equilibrating feedback mechanisms Marx examined.

These multiple future processes are locked in according to Lovelock since: 'there is no large negative feedback that would countervail temperature rise' that stem from various positive feedbacks (2006: 35). Monbiot succinctly expresses this: 'climate change begets climate change' (2006: 10, see Rial et al. 2004, for more technical analyses). With business as usual, the stock of greenhouse gases could treble by the end of this century. Indeed there is a 20 per cent risk of more than a five degrees centigrade increase in temperatures and the resulting transformation of the world's physical and human geography through the 5–20 per cent reduction in world consumption levels and significant decreases in global population (Stern 2006: iii, x, Stern 2007). According to the Stern Review, there are overwhelming and immediate economic reasons for *now* reducing global carbon consumption since there will be a 'high price to delay' (2006: xv). Especially significant is the potential melting of Greenland's ice cap, something that regularly happened in the relatively recent past with abrupt switches from cold to warm periods. This would dramatically change sea and land temperatures worldwide with the possible turning off of the Gulf Stream (Lovelock 2006: 33, Rial et al. 2004: 19). Pearce argues that the earth has not in the past changed temperatures gradually but dramatically, so that there has been very little time between glacial periods and periods of heating (2007).

Relatedly, oil supplies around the world are about to start running down. Peak oil production occurred in the US as far back as 1971 and it seems that oil production worldwide will peak around 2010 especially because of the failure to discover new fields at the same rate as in the past (Elliott and Urry 2010:

Chapter 6, Jackson 2006, Heinberg 2005, Leggett 2005). Transportation energy, which consists of at least 95 per cent oil, will be increasingly expensive and there will be frequent shortages especially with the world's population and its use of oil continuing to increase. There is not enough oil and this will generate significant economic downturns, more resource wars and lower population levels. Indeed the US seems to have based much of its foreign policy on the concept of peak oil for over 30 years and this policy accounts for the increased range and impact of Middle Eastern oil wars (Heinberg 2005). The delivery of fresh water also depends upon fossil fuels while severe water shortages face one third of the world's population (Laszlo 2006: 28f.). Rifkin claims that the oil age is 'winding down as fast as it revved up' (2002: 174).

Overall Rifkin argues: 'Like Rome, the industrial nations have now created a vast and complex technological and institutional infrastructure to sequester and harness energy' (2002: 62). This infrastructure was a twentieth century phenomenon with especially the US as *the* disproportionately high energy-consuming society. This resulted from its economy and society being based upon the specific combination of automobility *and* electricity. While the US possesses 5 per cent of the world's population it consumes a quarter of the world's energy and accounts for almost a quarter of global carbon emissions (Nye 1999: 6).

On Movement and Place

Much of the carbon use implicated in the 'American way of life' is generated by creating 'constructed' environments and by extensive growth in physical movement to and from such places. Especially important is long distant movement between different environments, as well as between family, friends and various places of tourist-type travel (see Larsen, Urry and Axhausen 2006). Carbon use within transport is the fastest growing source of greenhouse emissions, under the dominance of the US but also within Europe, Japan and Southeast Asia. In 1800 people in the US travelled 50 metres a day – they now travel 50 kilometres a day (Buchanan 2002: 121). Today world citizens move 23 billion kilometres; by 2050 it is predicted that this will increase fourfold to 106 billion if business continues as usual (Schafer and Victor 2000: 171).

Transport accounts for up to one-third of total carbon dioxide emissions (Geffen, Dooley and Kim 2003) and this stems from the many ways in which forms of life are now 'mobilized'. The growth of new kinds of long distance leisure, the establishment of globally significant themed environments, the growing significance of car and lorry transport within China and India, the rapid growth of cheap air travel, and the increased 'miles' flown and travelled by the world's 90,000 ships, by manufactured goods, foodstuffs and friends all exemplify such mobilized forms of life (see Larsen, Urry and Axhausen 2006, on friendship miles).

Central to changing climates is the reconfiguring of economy and society around 'mobilities' and 'constructed environments'. There is an emergent 'mobility complex', a new system of economy, society and resources that has been taking over the globe (Urry 2007). Large scale mobilities are not new but what is new is the development of this 'mobility complex' involving (see Elliott and Urry 2010, Urry 2007, 2008):

• the contemporary scale of movement around the world
• the diversity of mobility systems now in play
• the significance of the self-expanding automobility system and its risks
• the elaborate interconnections of physical movement and communications
• the development of mobility domains that by-pass national societies
• the significance of movement to contemporary governmentality
• constructed places of fascination and especially excess that have to be travelled to
• the development of a language of mobility, the capacity to compare and to contrast places from around the world
• an increased importance of multiple mobilities for people's social and emotional lives

Such a mobility field provides for the reconfiguring of emotional, physical and aesthetic conditions including new mobile lives and identities. Bauman argues that: 'Mobility climbs to the rank of the uppermost among the coveted values – and the freedom to move, perpetually a scarce and unequally distributed commodity, fast becomes the main stratifying factor of our late-modern or postmodern times' (1998: 2). Touring the world is how the world is increasingly performed with many people connoisseurs and collectors of places (especially aided by guides, web sites, TV programmes, word of mouth and so on). This connoisseurship, and hence the further amplification of mobility, applies to very many places, such as cities, landscapes, history, beaches, clubs, views, walks, surf, music scene, historic remains, sources of good jobs, food, landmark buildings, gay scene, party atmosphere, universities, museums, and so on. Included here are places of fascination and excess known for 'sex, drugs and rock and roll'.

Capitalist consumption in the 'rich north' towards the end of the twentieth century thus *escapes* from specific sites, as populations are mobile, moving in, across and beyond 'territories'. As people move around and develop personalized life projects through being freed from certain structures, so they extend and elaborate their consumption patterns *and* social networks. Many are less determined by site-specific structures, of class, family, age, career and especially propinquitous communities, as Giddens (1994) and Beck (1999) both argue. There are many new constructed places for performing 'meetingness' for those living mobile lives.

The site-specific disciplining of consumption becomes less significant with networking as a major component of consumerism. This spreads through social networking sites and increasing movement to new destinations. At least for the rich third of the world, partners, family and friends are a matter of choice increasingly spreading themselves around the world, together with their sites of meetings (as shown in Elliott and Urry 2010). There is a 'supermarket' of friends and acquaintances, and they depend upon an extensive array of interdependent systems of movement in order to connect with this distributed array of networks but often meeting up within distinct places, often of fascination.

Contemporary capitalism presupposes and generates some increasingly expressive bodies or habituses relatively detached from propinquitous family and community. They are emotional, pleasure-seeking and novelty-acquiring. Such bodies are on the move, able to buy and indulge new experiences located in new places and with new people. This develops what has been termed the 'experience economy' (Pine and Gilmore 1999). So as people escape the disciplinary confines of family and local community based on slow modes of travel, so they then encounter a huge array of companies comprising the experience economy. Capitalist societies involve new forms of pleasure, with many elements or aspects of the body being commodified (for those that can afford it). Expressive capitalism develops into a mobile and mobilizing capitalism with transformed, and on occasions, overindulged bodily habituses including that of multiple addictions. At the same time, many other people are employed in 'servicing' such habituses and they too are also often on the move (Lash and Urry 1994). There are many ways in which bodies are commodified in and through them moving about and being moved about.

What we might say then is that late twentieth century capitalism generated many places of fascination and this significantly has contributed to the high carbon consumption of the 'north', with some extension to the richer parts of the developing world (see cases in Cronin and Hetherington 2008). Thus life for some presupposes intermittent movement, with bodies flowing and intermittently encountering others in rich, face-to-face (and embodied) co-presence, especially in places of fascination. Some such places develop into what I will term places of excess capitalism. How did these in particular come to develop?

I use here the general distinction between societies of discipline and of control. In the former there was discipline realized within specific sites of confinement, such as the family, local community, school, prison, asylum, factory, clinic and so on (Foucault 1976, 1991, Goffman 1968). Individuals moved from one such site to another with each possessing its own laws and procedures. Surveillance was based upon fairly direct co-presence within that specific locus of power, within each 'local' panopticon. Power was internalized, face-to-face and localized within that site. And consumption was specific and regulated through each site, including the family and local community based upon slow modes of travel, except for the occasional annual break to places of temporary excess.

Seaside resorts in northern Europe in the late nineteenth century provided original places of excess (see Walton 2007, on *Riding on Rainbows* at Blackpool). These places of working class mass pleasure were premised upon a number of marked contrasts, between work and leisure, home and away, workspace and leisure space, and ordinary time and holiday time. Pleasure derived from these contrasts, for a week to be away from domestic and industrial routines and places. Such places provided a chance to 'let your hair down' for a week in a place of carnival before returning to normal (Shields 1991). And much of that excess was heavily regulated through the co-presence of one's family and to some degree one's community, who travelled at the same time to the same place. Being in such places was a source of aspiration for the emergent working class across Europe and North America. These disciplinary societies with their high levels of spatial and functional differentiation reached their peak in mid-twentieth century Europe and North America.

But over the twentieth century another system of power emerges, what Deleuze terms societies of control. Here power is fluid, de-centred and less site specific (Deleuze 1995, see Lacy 2005, on the history of the term). The sites of disciplinary confinement become less physically marked and many critiques of the effects of 'institutionalization' lead to the closure of former places of confinement. Many sets of social relations are not spatially internalized within specific sites. Gender relations are less confined to the family, work is both globalized and in part carried out in the home, schooling partly occurs within the media and so on. Consequently, states increasingly adopt complex control systems of recording, measuring and assessing populations that intermittently move, beginning with the system of the passport and now involving a 'digital order' able to track and trace individuals as they move about searching for new places of fascination and excess.

This assemblage induces new environments to be travelled to virtually and corporeally. In the control society, desire becomes generated through capitalism and this creates difference and movement. What then characterizes places of movement and excess? Such an assemblage consists of the following: there are distinct zones to be travelled to; gates control entry and exit; they are highly commercialized; many contain simulated environments; there is pleasure and fascination, no guilt; norms are unregulated by family/local community; they are sites of 'free choice'; there is liminal consumption; bodies and emotions are subject to commodification; some control at least is digitized; such places are globally known for their consumption/ pleasure excess; and these are sites of potential mass addiction.

The paradise beaches of the Mediterranean, the Caribbean, Australia and Thailand are a leading example of such places in an excess economy (Sheller 2009). Over the twentieth century these strange liminal places became highly desired, the beach signifying a symbolic 'other' to factories, work and domestic life. Near-naked bodies are be caressed by the sun with the global icon of the tanned body indicating the body as a 'mask' or 'sign'. And such beaches have become emblematic of excess, global icons of an ordering of the difference in temporality between work and leisureliness. Contemporary capitalism presupposes a consumption of *excess*, increasing aspects of the body being subject to intensive commodification within specific places of pleasure (Sheller and Urry 2004). The combination of mobility and expressive-ization, necessary for neo-liberal capitalist development, has helped to enhance the high carbon societies of the twentieth century.

Giddens describes such a process of increased mobility as involving increased 'freedom' since people are forced away from site-specific forms of surveillance to distanciated increasingly digital forms of control. Part of such a 'freedom' is the freedom to become 'addicted', to be emotionally and/or physically dependent upon certain products of global capitalism, whether legal, illegal or semi-legal. So places of excess can be seen where significant numbers can be addicted and places of excess are synonymous with places of significant addiction. Giddens recently suggested that compulsive behaviour is so common in modern society because it is:

> linked to lifestyle choice. We are freer now than 40 years ago to decide how to live our lives. Greater autonomy means the chance of more freedom. The other side of that freedom, however, is the risk of addiction. The rise of eating disorders coincided with the advent of supermarket development in the 1960s. Food became available without regard to season and in great variety, even to those with few resources (Giddens 2007).

Addictions have a common basis in compulsive repetition – habits that are hard to break because of their emotional significance. In the contemporary world, many industries are premised upon addiction: tobacco, 'illegal' drugs, alcohol, gambling and sugar. The last of these is central to the generation of global obesity, a specific version of global excess. According to the Chair of the UK's National Obesity Forum, the health threat posed by obesity; 'will hit us much earlier than climate change ... We are now in a situation where levels of childhood obesity will lead to the first cut in life expectancy for 200 years. These children are likely to die before their parents.' (see BBC News 2007). Obesity is realized in many places of fascination, in food plazas, fast food chains, attractive restaurants, and places of promoted taste.

More generally, during this neo-liberal period new or designed places of consumption excess developed, involving various 'dispossessions', of workers' rights, of peasant land-holdings, of the state's role in leisure, of neighbourhood organizations, and of customary rights. David and Monk refer to these resulting places as 'evil paradises' and examples include Arg-e Jadid, a Californian oasis in the Iranian desert; the buildings for the $40 billion 2008 Olympics in Beijing, Palm Springs gated community in Hong Kong; Sandton in Johannesburg; and Macao (Davis and Monk 2007a, 2007b, see Dear's chapter in this book). The last of these involves a $25 billion investment oriented to providing leisured gambling for the reinvented China and the resurgent 1.3 billion Chinese. Simpson notes that in the very same year that Macao was declared a Unesco World Heritage site the first steps occurred in constructing a vast 'Fisherman's Wharf' of themed reproductions of a Roman Coliseum, buildings from Amsterdam, Lisbon, Cape Town and Miami, and an exploding volcano (see Simpson's chapter in this book). Globally the next planned paradise is the €17 billion entertainment city Gran Scala in north east Spain. There are 32 planned casinos, 70 hotels, 232 restaurants, 500 stores, a golf course, a race track and a bullring, located of all places in the Los Monegros desert, where water and oil are overwhelming resource issues.

Consuming the Planet

This reconfiguration of environments through the excess economy has severe environmental consequences. Indeed, it is those very same beaches at tourist destinations of paradise that will be some of the very first settlements to disappear under rising sea levels and increased extreme weather events due to greenhouse gas emissions in part caused by the planes bringing in those tourists (Amelung, Nicholls and Viner 2007). Consumption thus turns out to involve the profligate consumption of resources. Resources have bitten back.

Building for the Olympics in Beijing involved constructing the equivalent of three Manhattans; one-half of the world production of concrete and one-third of its steel is being used more generally to construct 'urban China'. And in this pantheon of excess Dubai is surely number one. It is said to have been the world's largest building site with dozens of megaprojects including the palm islands, the 'island world', the world's only seven star hotel, a domed ski resort where sub-zero temperatures are maintained in the middle of a baking desert, and carnivorous dinosaurs (Davis 2007). This is a place of monumental excess, where being number 1 in the world is the ambition. If Dubai is to be the luxury-consumer paradise of the Middle East and South Asia: 'it must ceaselessly strive for visual and environmental excess' (Davis 2007: 52). And it achieves this through architectural gigantism and perfectibility with simulacra more perfect than the original. The UAE ranks second in the global league table of per capita carbon emissions (beaten only by its Gulf neighbour, Qatar). Dubai might be described then as a vast gated community of excess where state and private enterprises are virtually indistinguishable (Davis 2007: 61). It is a place of vice, of overconsumption, prostitution, drink and gambling where guilt is not based on consuming to the 'limit'.

And this is all made possible by migrant contract labourers from Pakistan and India bound to a single employer and subject to totalitarian controls. Almost all of the labour is imported and passports are removed from migrants on entry. Over a million men and women from India, Bangladesh, Nepal and across Asia have turned Dubai from a sleepy village of pearl-divers and fishermen into a shimmering Arabian Las Vegas and have been rewarded with next to no rights and meagre pay. They sleep in labour camps, each one crammed with 3,000 or more people. In the strict hierarchy of the emirate, their role is to serve the expats and wealthy natives.

But the hubris that was Dubai seems to have gone into very rapid reverse. Dubai is a perfect metaphor for the crisis crippling global capitalism since 2008. The dream of Sheikh Mohammed bin Rashid al-Maktoum, was unsustainable economically and environmentally. Its astonishing growth has reversed. Expats are fleeing and leaving their cars bought on credit at the airport, thousands of construction workers have been laid off, there is a predicted 60 per cent fall in property values, half the construction projects are on hold or cancelled, the population is now shrinking and Dubai World, its major company, is unable to meet its debts. Jon Henley summarizes Dubai: 'You can see it as an utterly unsustainable abomination of greed and excess, built in the middle of an inhospitable and bone-dry desert with what amounts to slave labour at who knows what cost to the environment (one golf course alone consumes 4m gallons of desalinated water a day; the carbon footprint of the average Dubai resident is more than double even the average American's)' (Henley 2009, and see Davidson 2008, on the vulnerability of Dubai's success). Is this history of the present in Dubai a forerunner of the history of the present much more generally as mobile lives possibly come to a shuddering halt within the next few decades?

Conclusion

In the end, the many interdependent systems of fascination and excess are based upon high and expanding resource use. It is these economies and segmented societies of excess that are moving 'the world' in a seemingly inexorable fashion towards an abyss. As Davis and Monk argue:

> On a planet where more than 2 billion people subsist on two dollars or less a day, these dreamworlds enflame desires – for infinite consumption, total social exclusion and physical security, and architectural monumentality – that are clearly incompatible with the ecological and moral survival of humanity (Davis and Monk 2007b: xv).

Only some exceptionally powerful systems could offset this contradiction between high and growing levels of consumption and its global carbon impact. And there is only limited time to do so. According to the economist Stern: 'Climate change ... is the greatest and widest-ranging market failure' stemming from the most intense period of 'neo-liberalism' (2006: i). The adaptive and evolving relationships between enormously powerful systems especially those of a de-regulated economy are, as Giddens once expressed it via a mobility analogy, a 'juggernaut' careering at full pace to the edge of the cliff (1990, and see 2009, on climate change and 'policies').

The overall economic, social and political consequences of such unique changes are global and, if they are not significantly mitigated, will very substantially reduce the standard of living and population worldwide – especially since these impacts will put an even stronger pressure and impact upon poorer countries in Africa and parts of Asia (Roberts and Parks 2007, Stern 2007).

Thus if global heating continues to escalate through positive feedback loops then 'regional warlordism' might be a likely scenario for 2050 (Urry 2007: Chapter

13, Foresight 2006, Kunstler 2006). This would involve the substantial breakdown of many mobility, energy and communication connections currently straddling the world. There would be a plummeting standard of living, a relocalization of mobility patterns, an increasing emphasis upon local warlords controlling recycled forms of mobility and weaponry, and relatively weak imperial or national forms of governance. There would be an increasing separation between different regions, or 'tribes'. And, as noted, the first places to disappear would be paradise beaches, islands and other coastal settlements of excess. September 2005 New Orleans shows what this scenario would be like for a major city in the rich but unequal 'north' (Hannam, Sheller and Urry 2006). Systems of repair would dissolve with localized recycling of bikes, cars, trucks, computer and phone systems. Only the super-rich would travel far and they would do so in the air within armed helicopters or light aircraft, with very occasional tourist-type space trips to escape the hell on earth in space as the new place of excess. Those who could find security in gated and armed encampments would survive, with the privatizing of collective functions and the further gating of pleasure zones combining fascination and fear in a new way.

There could well be a Hobbesian war of warlords, where defence becomes a necessity against their neighbours, especially for control of water, oil and gas. And with extensive flooding, extreme weather events and the break-up of long distance oil and gas pipelines, these resources would be exceptionally contested and defended by armed gangs. There would be no monopoly of physical coercion in the hands of national states, little disciplinary power. Life even in the 'north' would be nasty, brutish and almost certainly 'shorter'. Life in parts of the 'south' is already being transformed in such a way by global climate change. And there have been various oil wars. The heated-up planet will also usher in water wars; it is calculated that a temperature increase of 2.1 degrees would expose up to 3 billion people to water shortages (Monbiot 2006: 6). Other 'wild zones' will develop where the 'north' will *exit* as fast as possible, if and when the oil or water no longer seem to flow especially through the places of excess (as they exited from New Orleans in 2005). Bauman notes how one key element of contemporary power is 'escape, slippage, elision and avoidance, the effective rejection of any territorial confinement' and the possibility of escape into 'sheer inaccessibility' increasingly through 'offshoring' (Bauman 2000: 11). There are many examples of such 'exitability' for the kinetic elite who function as 'absentee landlords' with high mobility and high potential for exit mobility if the 'going gets tough' (Bauman 2000: 13). Societies in the poor south, including various zones of excess, will be exited from and left to ethnic, tribal or religious warlordism of a traditional kind, with multitudes from time to time re-entering the safe zones as migrants working in the security-ized zones of excess or as slaves or as terrorists (Bauman 2000, Lash and Urry 1994).

Late neo-liberal, twentieth century capitalism has generated the most striking of contradictions. Its pervasive, mobile and promiscuous commodification involved utterly unprecedented levels of energy production and consumption, a high carbon society whose dark legacy we are beginning to reap. Fascination has turned out to have profound consequences, both environmentally and socially. This could result in a widespread reversal of many of the systems that constitute capitalism as it becomes its own gravedigger. As Davis and Monk apocalyptically argue: 'the indoor ski slopes of Dubai and private bison herds of Ted Turner represent the ruse of reason by which the neoliberal order both acknowledges and dismisses the fact that the current trajectory of human existence is unsustainable' (2007b: xvi). Excess pleasure during neo-liberalism would seem to have generated an excess potential of catastrophe in our new but not so fascinating twenty-first century (see Elliott and Urry 2010, Chapter 7, for alternative scenarios for 2050).

Acknowledgements

This chapter is drawn from Elliott and Urry 2010. I am grateful for these various collaborations with Anthony Elliott, Heiko Schmid and Woody Sahr.

References

Amelung, B., Nicholls, S. and Viner, D. 2007. 'Implications of Global Climate Change for Tourism Flows and Seasonality'. *Journal of Travel Research*, 45(3), 285–96.
Bauman, Z. 1998. *Globalization: The Human Consequences*. Cambridge: Polity Press.
——— 2000. *Liquid Modernity*. Cambridge: Polity Press.
BBC News 2007. 'Obesity "as bad as climate risk"'. *BBC News* [Online, 14 October]. Available at: http://news.bbc.co.uk/1/hi/health/7043639.stm. [accessed: 15 October 2007].
Beck, U. 1999. *Individualization*. London: Sage.
Beinhocker, E.D. 2006. *The Origin of Wealth: Evolution, Complexity, and the Radical Remaking of Economics*. London: Random House.
Buchanan, M. 2002. *Nexus: Small Worlds and the Groundbreaking Science of Networks*. London: W.W. Norton.
Cilliers, P. 1998. *Complexity and Postmodernism: Understanding Complex Systems*. London: Routledge.
Cronin, A. and Hetherington, K. (eds) 2008. *Consuming the Entrepreneurial City: Image, Memory, Spectacle*. London: Routledge.
Davidson, C. 2008. *Dubai. The Vulnerability of Success*. London: Hurst and Company.
Davis, M. 2007. 'Sand, Fear, and Money in Dubai', in *Evil Paradises. Dreamworlds of Neoliberalism*, edited by M. Davis and D. Monk. New York: The New Press, 48–68.

Davis, M. and Monk, D. (eds) 2007a. *Evil Paradises. Dreamworlds of Neoliberalism*. New York: The New Press.
———— 2007b. 'Introduction', in *Evil Paradises. Dreamworlds of Neoliberalism*, edited by M. Davis and D. Monk. New York: The New Press, ix–xvi.
Deleuze, G. 1995. *Negotiations, 1972–1990*. New York: Columbia University Press.
Elliott, A. and Urry, J. 2010. *Mobile Lives*. London: Routledge.
Foresight 2006. *Intelligent Information Futures. Project Overview*. London: Department for Trade and Industry.
Foucault, M. 1976. *The Birth of the Clinic*. London: Tavistock.
———— 1991. 'Governmentality', in *The Foucault Effect. Studies in Governmentality*, edited by G. Burchell, C. Gordon and P. Miller. London: Harvester Wheatsheaf, 87–104.
Geffen, C., Dooley, J. and Kim, S. 2003. *Global Climate Change and the Transportation Sector: an Update on Issues and Mitigation Options*. Paper to the 9th Conference: Diesel Engine Emission Reduction Conference, USA.
Giddens, A. 1990. *The Consequences of Modernity*. Cambridge: Polity Press.
———— 1994. 'Living in a Post-traditional Society', in *Reflexive Modernization: Politics, Tradition and Aesthetics in the Modern Social Order*, edited by U. Beck, A. Giddens and S. Lash. Cambridge: Polity Press, 56–109.
———— 2007. 'All Addictions Turn from Pleasure to Dependency'. *The Guardian* [Online: 16 October]. Available at: http://www.guardian.co.uk/comment/story/0,,2191886,00.html [accessed: 22 November 2007].
———— 2009. *The Politics of Climate Change*. Cambridge: Polity Press.
Goffman, E. 1968. *Asylums: Essays on the Social Situation of Mental Patients and other Inmates*. London: Penguin.
Hannam, K., Sheller, M. and Urry, J. 2006. 'Editorial: Mobilities, Immobilities and Moorings'. *Mobilities*, 1(1), 1–22.
Heinberg, R. 2005. *The Party's Over: Oil, War and the Fate of Industrial Society*. New York: Clearview Books.
Henley, J. 2009. 'Dubai: Business as Usual'. *The Guardian* [Online, 3 December]. Available at: http://www.guardian.co.uk/business/2009/dec/03/dubai-world-economic-crash [accessed: 7 December 2009].
IPCC, Intergovernmental Panel on Climate Change (ed.) 2007. *Climate Change 2007: Synthesis Report. Contribution of Working Groups I, II and III to the Fourth Assessment Report of the Intergovernmental Panel on Climate Change*. Geneva: IPCC.
Jackson, P. 2006. *Why the Peak Oil Theory Falls Down: Myths, Legends, and the Future of Oil Resources*. [Online, 10 November]. Available at: http://www.cera.com/aspx/cda/client/report/reportpreview.aspx?CID=8437&KID= [accessed: 5 July 2010].
Kunstler, J.H. 2006. *The Long Emergency: Surviving the Converging Catastrophes of the Twenty-First Century*. London: Atlantic Books.
Lacy, M. 2005. *Security and Climate Change*. London: Routledge.

Larsen, J., Urry, J. and Axhausen, K. 2006. *Mobilities, Networks, Geographies*. Aldershot: Ashgate.

Lash, S. and Urry, J. 1994. *Economies of Signs and Space*. London: Sage.

Laszlo, E. 2006. *The Chaos Point*. London: Piatkus Books.

Leggett, J. 2005. *Half Gone. Oil, Gas, Hot Air and Global Energy Crisis*. London: Portobello Books.

Lovelock, J. 2006. *The Revenge of Gaia: Earth's Climate Crisis and the Fate of Humanity*. London: Allen Lane.

Lynas, M. 2007. *Six Degrees. Our Future on a Hotter Planet*. London: Fourth Estate.

Marx, K. and Engels, F. 1888 [1848]. *The Manifesto of the Communist Party*. Moscow: Foreign Languages.

Maynard Smith, J. 1994. 'Comments', in *Complexity, Metaphors, Models and Reality*. Studies in the Sciences of Complexity Proceedings, vol. 19, edited by G. Cowan, D. Pines and D. Meltzer. Santa Fe: Santa Fe Institute.

Monbiot, G. 2006. *Heat. How to Stop the Planet Burning*. London: Allen Lane.

Nye, D.E. 1999. *Consuming Power*. Cambridge, MA: MIT Press.

Pearce, F. 2007. *With Speed and Violence. Why Scientists fear Tipping Points in Climate Change*. Boston: Beacon Press.

Pine, B.J. and Gilmore, J. 1999. *The Experience Economy*. Cambridge: Harvard Business School Press.

Rial, J. et al. 2004. 'Nonlinearities, Feedbacks and Critical Thresholds within the Earth's Climate System'. *Climate Change*, 65(1–2), 11–38.

Rifkin, J. 2002. *The Hydrogen Economy*. New York: Penguin Putnam.

Roberts, J.T. and Parks, B.C. 2007. *A Climate of Injustice: Global Inequality, North-South Politics, and Climate Policy*. Cambridge, MA: MIT Press.

Schafer, A. and Victor, D. 2000. 'The Future Mobility of the World Population'. *Transportation Research A*, 34(3), 171–205.

Sheller, M. 2009. 'The New Caribbean Complexity: Mobility Systems, Tourism and Spatial Rescaling'. *Singapore Journal of Tropical Geography*, 30(2), 189–203.

Sheller, M. and Urry, J. (eds) 2004. *Tourism Mobilities*. London: Routledge.

Shields, R. 1991. *Places on the Margin: Alternative Geographies of Modernity*. London: Routledge.

Stern, N. 2006. *Stern Review. The Economics of Climate Change*. [Online]. Available at: http://www.hm-treasury.gov.uk/independent_reviews/stern_review_economics_climate_change/sternreview_index.cfm [accessed: 6 November 2006].

————— 2007. *The Economics of Climate Change. The Stern Review*. Cambridge: Cambridge University Press.

Szerszynski, B. and Urry, J. (eds) 2010. *Changing Climates. Special Issue of Theory, Culture and Society*, 27(2–3), 1–305.

Urry, J. 2007. *Mobilities*. Cambridge: Polity Press.

————— 2008. 'Climate Change, Travel and Complex Futures'. *British Journal of Sociology*, 59(2), 261–79.

Walton, J. 2007. *Riding on Rainbows: Blackpool Pleasure Beach and its Place in British Popular Culture*. St Alban's: Skelter.

Index

theories of the state 101
state effects *see* effects
state power *see* power
stimulus 7, 56, 149, 152
strategy 10–11, 26, 44, 61, 81, 84, 98–100,
    102–3, 111, 116, 147–51, 156,
    165–6, 171, 173, 176, 179, 181–4,
    187, 195, 203
    image/imagery strategy 11
    marketing strategies 84, 169, 171,
        173–4, 176, 181, 183
    planning strategies 10, 122
    promotion strategies 108, 149, 170
structuration 110, 117–18, 120, 122
    social structuration 8, 9
    political structuration 6
structure 6, 37, 56, 79–80, 86, 110,116–19,
    121–2, 138, 144, 151, 156, 158,
    160–63, 165, 180, 213
    administrative structure 119
    power structures 7, 51, 84, 141
    spatial structure 19, 76, 79
    urban structure 12, 19
studies
    language studies 133
    urban studies 8, 43, 79
subject 6, 10, 47, 51–2, 56, 63, 97, 110,
    114–17, 147, 192, 195, 198–200,
    202–3
    neoliberal subject 191, 193, 198
subjectivity 8, 10–12, 58, 71, 107, 109,
    111, 113, 115, 117–20, 122–3, 199
    urban subjectivities 1, 119–20
suburban periphery *see* periphery
suburbanization 17, 23–4
suburbia 80, 82, 84, 86
suburbs 18, 22–3, 142
surplus of meaning *see* meaning
surrounding environment *see* environment
surrounding reality *see* Herumwirklichkeit
surveillance 27, 99, 214, 216
    security and surveillance *see* security
sustainability 20, 23, 66–7
    environmental sustainability 24
symbolic economy *see* economy
symbolic process *see* process
symbolic value *see* value
symbolism 64, 66

system
    bus system 120–21
    capitalist system 6, 210
    knowledge system 173, 176
    non-equilibrium system 12, 210
    value system 17

table game *see* game
telecommunication 38, 43
territorialization 110, 112
territory 96, 99, 101, 115, 117, 122, 150,
    158, 165, 180, 188–9, 200, 213
    Palestinian territories 99
terror 5, 115
terrorism 99, 103
The Three Princes of Serendip 44
themed casino *see* casino
theming 1–2, 8, 81, 84, 87, 112
theory 4, 10, 17, 19–21, 26, 28, 36, 56,
    62, 64, 68–70, 75–7, 81, 88, 97–8,
    100–101, 107, 112, 123, 130, 133,
    138, 160
    action theory 56, 62, 138
    cultural theory 96
    discourse theory 173
    Marxist theory 38
    post-structuralist theory 96–7, 102
    social theory 19, 37, 47, 68, 96, 110
    urban theory 8, 19, 24–5, 44, 75, 133
    The Theory of the Leisure Class 1
thing 108, 122
thinking
    mode of thinking 68
    mood of thinking 68–9
    relational thinking 109
thought
    modernist thought 21
    postmodern thought 20
    urban thought 37
A Thousand Plateaus 108, 117
thrill of gratification and seduction 6
Tokyo 45
touching gaze *see* gaze
tourism 18, 147–8, 153–4, 161–2, 179,
    182, 188–9, 192, 198
tourist gaze *see* gaze
town
    landscape of towns 35